INSIGHT
BIOLOGY

Theory of Practicals for
SSCE/GCE/NECO Candidates

100
Questions and
Answers

Compiled by:

ADEWOYE B. A.

NEWMAN SPRINGS PUBLISHING
320 Broad Street
Red Bank, NJ 07701

First originally published by Newman Springs Publishing 2022

ISBN 978-1-68498-180-9 (Paperback)
ISBN 978-1-68498-181-6 (Digital)

Printed in the United States of America

To the only one God, whose thoughts are not our
thoughts, neither are our ways His ways

CONTENTS

Foreword..vii

Acknowledgments ix

Questions...1

Answers...175

FOREWORD

Insight Biology: Theory of Practicals for SSCE/GCE/NECO Candidates is a comprehensive attempt to maximally expose candidates offering biology to the practical exam as it is normally set. It consists of one hundred carefully designed model questions and model answers to ensure that candidates are adequately exposed to practical work in biology.

The questions span through the senior school certificate curriculum in biology and include the practical specimens that have been set in the past and those that may possibly be set. Every specimen has been appropriately drawn in such a way as to make identification easy.

The model answers are also very thorough so that candidates are adequately guided as to the correct response to every question asked.

School candidates offering biology at the senior-school certificate level and external candidates who have to do the alternative to practical exam in biology will find the book very useful.

The author has painstakingly provided a textbook that will adequately prepare candidates for practical work in biology. I strongly recommend the text for serious-minded students who have the ambition to have the best in their academic pursuit. It is also a valuable text for teachers of biology at the senior-school certificate level. It is indeed the text students have been waiting for.

S. K. Olalere
BSc (Hons) PGDE
Retired Deputy Director of Education (Biology)
Federal Ministry of Education

ACKNOWLEDGMENTS

The Lord be praised for his mercies, love, strength, and all adorable virtues bestowed on me that helped in the compilation of this textbook.

The West African Examinations Council is duly acknowledged for delving into its archives and making use of its past questions that were seriously amended to satisfy the author's style. Diagrams used were from various textbooks whose authors are hereby acknowledged.

I cherish the love and concern of my parents, late Professor Sam. A. Adewoye (retired) and Mrs. A. A. Adewoye (retired), both of University of Ilorin, English Department, and University Library respectively.

Permit me to say, glory be to God in the highest for the past that creates way for a turning-around present, and with God in control, a glorious future is certain.

Your academic lives have been a challenge and source of inspiration. Mr. S. K. Olalere, who wrote the foreword, is a man to be highly acknowledged. He has succeeded in putting me in the right path in the teaching of biology, and it is to be noted that this book is a by-product of his tutelage.

To my friends and colleagues who helped in the editing and proofreading, Mr. A. O. Atanda (Federal Ministry of Education, Nigeria) and Mr. Shola Ojo, also a brother (now in the United Kingdom), I say I am very grateful for sparing your precious time.

Another person that needs to be acknowledged is Mr. E. U. Chuckwuka (Federal Ministry of Education, Nigeria), a man whose library I used judiciously in his absence. Other colleagues of mine in the field of biology education are equally acknowledged: Mrs. D. O. Oderinde and Mrs. V. O. Taiwo.

To my siblings, I always appreciate your love.

All those that worked on the manuscripts, too numerous to mention, are also appreciated.

To my wife, Grace Olufunsho, I say a big thank-you for your encouragement and for being there at the home front. My children—Israel OreOluwa; Zion Toluwalomo; Lois Mofoluwato; and Emmanuel Obaloluwa, our surprise baby—I love you all.

To all who shed light during the compilation, thank you unalloyedly.

QUESTIONS

1A. Identify the organisms below:

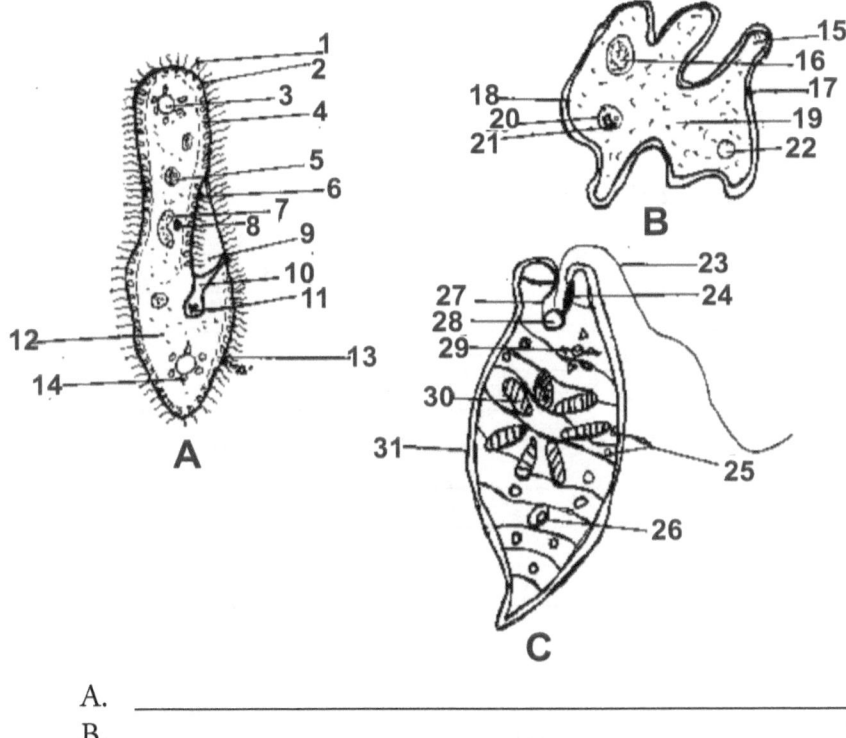

A
B
C

A. _____

B. _____

C. _____

1B. Give the names of the structures labeled: 1–31.

1. _____   12. _____

2. _____   13. _____

3. _____   14. _____

4. _____   15. _____

5. _____   16. _____

6. _____   17. _____

7. _____   18. _____

8. _____   19. _____

9. _____   20. _____

10. _____   21. _____

11. _____   22. _____

23. _____   28. _____
24. _____   29. _____
25. _____   30. _____
26. _____   31. _____
27. _____

1C. Compare the three organisms using the table given.

	A	B	C
Shape			
Food capture			
Site of food digestion			
Absorption of food			
Respiration			
Excretion			
Reproduction			
Sensitivity			
Movement			

1D. Give the functions of the parts labeled: 6, 13, 20, 22, and 23.

Functions

Part labeled 6: _____

Part labeled 13: _____

Part labeled 20: _____

Part labeled 22: _____

Part labeled 23: _____

1E. Give the phylum to which the three organisms belong to:

_____

1F. Why is the organism labeled C regarded as a plant?

_____

_____

_____

1G. Where can the organisms labeled A, B, and C be found?

A. _____

B. _____

C. _____

2A. *Study the diagram below and use it to answer questions 2A (i–iv).*

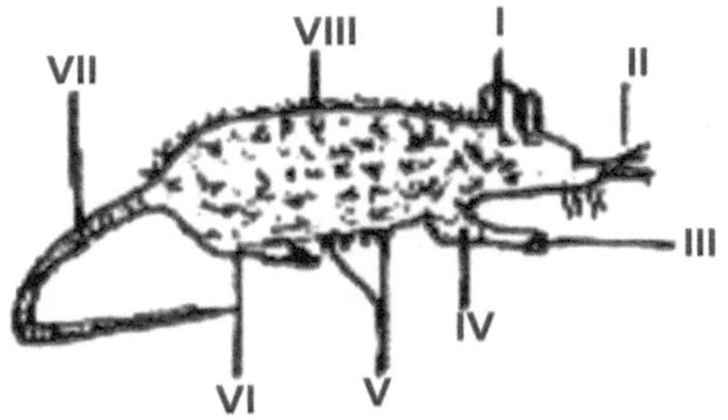

i.   Identify the organism illustrated in the diagram.

_____

ii. Name the parts labeled: I–VIII.

I.   _____   III. _____

II. _____   IV. _____

V. _____   VII. _____
VI. _____   VIII._____

iii. List three similar and two different observable features between man and the organism illustrated in the diagram above.

Three similar observable features

   I.   _____
   II. _____
   III. _____

Two different observable features

   I.   _____
   II. _____

iv. State the function of the parts labeled (II and V):

   II. _____
   V. _____

3A. *Study the diagram below and use it to answer question 3.*

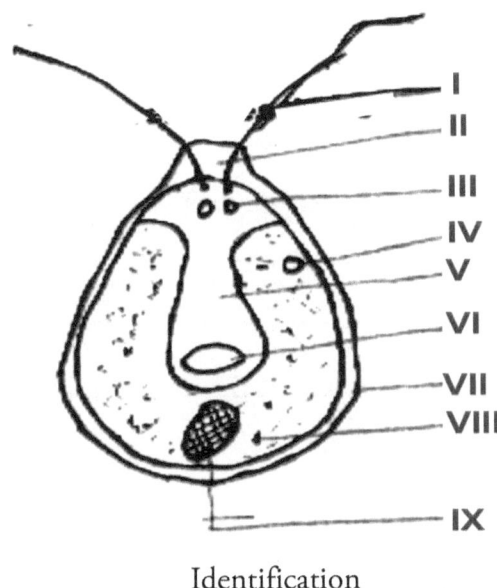

Identification

i.   Identify the organism illustrated in the diagram:

_____

ii. Name the parts labeled I to VIII:

I.   _____

II.   _____

III. _____

IV.   _____

V.   _____

VI.   _____

VII. _____

VIII. _____

3B.

    i.   Is the organism a plant or an animal?

    _____

    ii. Give the reasons for your answer above:

    _____
    _____

3C. Give one function of each of the structure labeled III, IV and VI.

    Functions:

        III. _____
        IV. _____
        VI. _____

3D. Why is the organism regarded as primitive?

Diagrams I and II illustrate the alimentary systems of bird and cockroach. Use them to answer questions 4A to 4D.

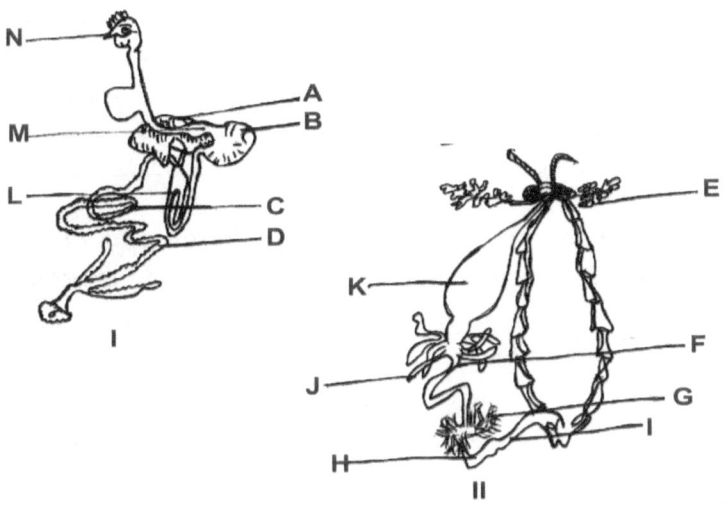

4A. Name the structures labeled A to N.

A. _____

B. _____

C. _____

D. _____

E. _____

F. _____

G. _____

H. _____

I. _____

J. _____

K. _____

L. _____

M. _____

N. _____

4B. List *one function* and *one adaptive feature* of each of the structures labeled B, J, K, and M.

B. _____

J. _____

K. _____

L. _____

M. _____

4C. Using the diagrams, list *two similarities* and *three differences* between the alimentary systems of bird and cockroach.

Two similarities:

i. _____

ii. _____

Three differences:

i. _____

ii. _____

iii. _____

4D.

i. What type of food would the animal represented by Diagram I feed on?

ii. Give two reasons for your answer:

1. _____

2. _____

Study the diagram below and use it to answer questions A through E.

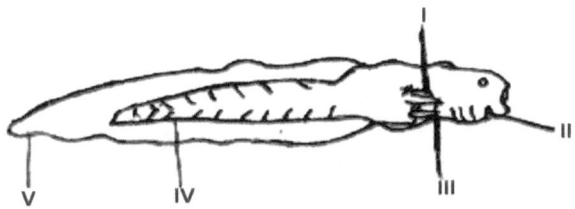

5A. Identify the organism illustrated in the diagram:

_____

5B. Name the parts labeled (I–V) in the diagram.

I. _____
II. _____
III. _____
IV. _____
V. _____

5C. What is the function of the part labeled I.?

I. _____

5D. Name the habitat of that organism.

_____

5E. How does the organism:

i. Feed? _____
ii. Move? _____

Diagrams I and II represent stages in the development of a toad. Study them carefully and use them to answer questions A to G.

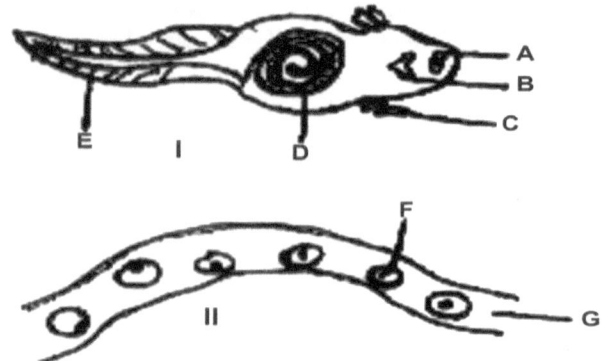

6A. Name the structures labeled A to E:

A. _____   D. _____
B. _____   E. _____
C. _____

6B. What are the functions of the parts labeled A and C?

A. _____
C. _____

6C. Name the next stage after the stage labeled I.

_____

6D. Name the parts labeled F and G in diagram II.

F. _____
G. _____

6E. List two functions of the part labeled G.

(i) _____

(ii) _____

6F. Briefly explain the relationship between diagrams I and II.

_____

_____

6G. State one adaptive feature of the structure labeled F.

_____

The diagram below illustrates two homologous chromosomes during meiosis. Use it to answer question 7.

7A. Name the point labeled I:_____

7B. What is the importance of those points in evolution?

_____

7C. What is this state of meiosis called?

_____

Study the diagrams below and use them to answer questions A to D.

8A. Which method of reproduction is illustrated in diagrams A and B?

_____

8B. Give the specific names of the types of reproduction illustrated in diagrams A and B.

A. _____

B. _____

8C. Give an example each of organisms that reproduce by the methods illustrated in diagrams A, B, and C.

A. _____

B. _____

C. _____

8D. State three (i) advantages and three (ii) disadvantages of the method of reproduction named in 8A.

Advantages:

i. _____
ii. _____
iii. _____

Disadvantages:

i. _____
ii. _____
iii. _____

Study the diagram below and use it to answer questions A to E.

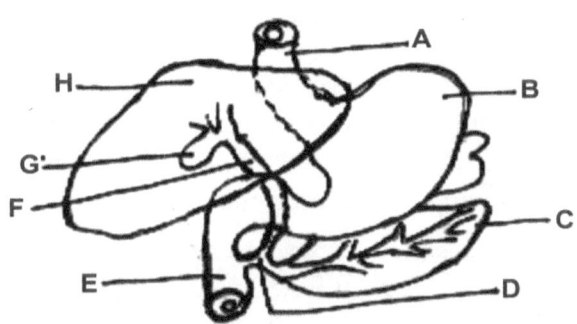

10A. Name the parts labelled A to H in the diagram.

A. _____   H. _____
B. _____
C. _____
D. _____
E. _____
F. _____
G. _____

10B. In which of the labeled parts is bile secreted?

_____

10C. Which of the labeled parts secretes both hormones and digestive juice?

_____

10D. State the hormone secreted by the organ mentioned above (9C):_____

10E. How does the malfunctioning of the part labeled H affect digestion in man?

_____

_____

The data below were recorded from the tests carried out on water collected from two points, A and B, of a flowing river where sewage had been discharged.

Use the following information to answer questions A to D:

Conditions	Point A	Point B
Oxygen	16 ppm	8 ppm
Carbon dioxide	19 ppm	27 ppm
pH	9.9	9.0
Organisms	Higher-population density	Very low population density

(*ppm* means "parts per million")

10A. Explain the difference in oxygen and carbon dioxide concentrations at points A and B:

At point A:_____

At point B:_____

10B. Explain the difference in pH at point A and B:

At point A:_____

At point B:_____

10C. Why are there more organisms at point A than at point B?

_____

_____

10D. State the type or organisms that can be found at point B.

_____

_____

Diagrams I and II below represent pitcher plant and tapeworm respectively.

Study the diagrams carefully and use them to answer questions A to F.

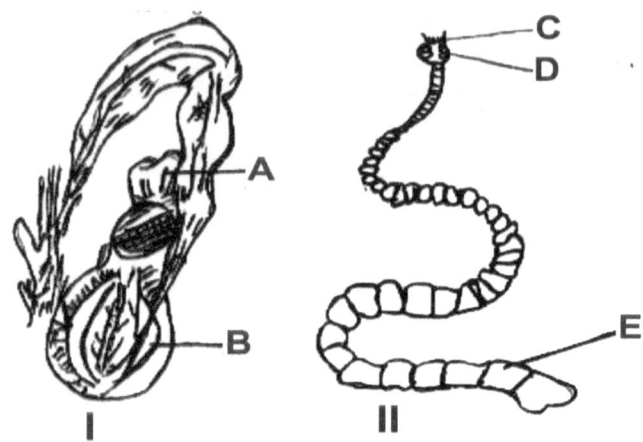

11A. Identify the structures labeled A to E.

A. _____ D. _____

B. _____ E. _____

C. _____

11B. State the mode of feeding of the organisms illustrated in diagrams I and II:

I. _____

II. _____

11C. State two ways in which the mode of feeding of tapeworm differs from the mode of feeding of Rhizopus.

Tapeworm	Rhizopus
(i) _____	_____
(ii) _____	_____

11D. State the food substance which the organism represented in diagram I feeds on:

_____

11E. State two structural features of the organism represented in diagram II which adapt it to its mode of life:

(i). _____

(ii). _____

11F. Briefly describe the mechanism of feeding of the organisms illustrated in diagrams I and II.

Diagram I:_____

Diagram II:_____

11G. What is the habitat of the adult tapeworm?

_____

11H. State two effects of tapeworm infection on man:

(i). _____

(ii). _____

Study diagrams I and II and use them to answer question 12.

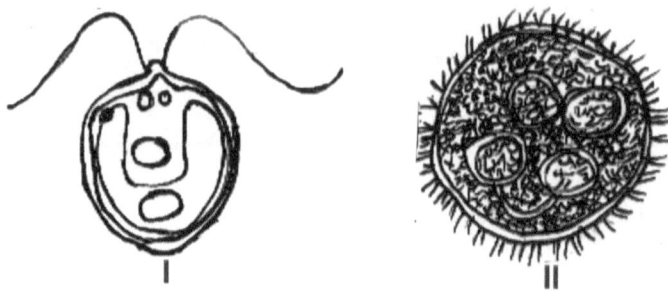

12A. Identify the organisms illustrated in diagrams I and II:

    I. _____

    II. _____

12B. State the levels of organizations in diagram I and II:

    I. _____

    II. _____

12C. State a reason for placing I and II in that level of organization:

    I. _____

    II. _____

The diagrams below illustrate certain features of two types of fruits. Use them to answer questions A through F.

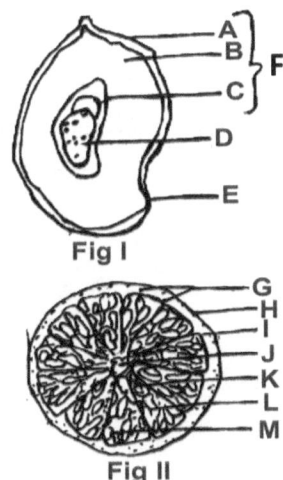

Fig I

Fig II

13A. What view is shown in:

   i.   Fig. I?_____

   ii. Fig. II?_____

13B. Name the parts labeled A to F in Fig. I:

   A. _____

   B. _____

   C. _____

   D. _____

   E. _____

   F. _____

13C. List (i) three differences and (ii) one similarity between the two types of fruits shown in Fig. I and Fig. II:

Differences:

i.) _____
ii.) _____
iii.) _____

Similarity:

_____

13D. Give two differences between Fig. I and a typical seed:

Differences:

I. _____

_____

II. _____

_____

13E. Describe one method of dispersal for the fruits illustrated in Fig. II:

_____

_____

13F. Name the parts labeled G to M:

G. _____
H. _____
I. _____
J. _____
K. _____
L. _____
M. _____

The diagram below illustrates an ecosystem. Use it to answer questions A through G.

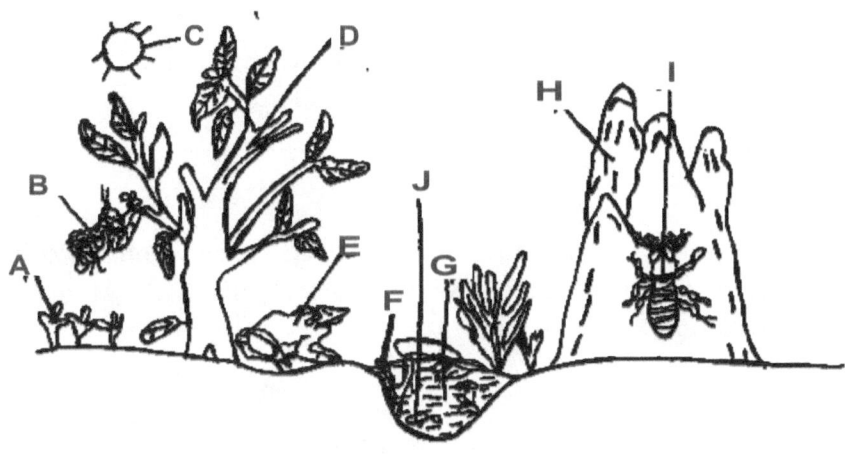

14A. Name the ecosystem illustrated in this diagram:

_____

14B. Name the components labeled A to J in the diagram:

A. _____

B. _____

C. _____

D. _____

E. _____

F. _____

G. _____

H. _____

I. _____

J. _____

14C. Name any two consumers in the ecosystem illustrated in the diagram:_____

14D. Describe the energy transfer within this ecosystem.

_____

_____

14E. If a hawk is introduced into the ecosystem, construct in the space provided below a food web involving as many components as possible.

14F. Suggest any two decomposers within this ecosystem?

_____

_____

14G. What is the climax vegetation in this ecosystem?

_____

_____

Study the diagram below and use it to answer questions A through F.

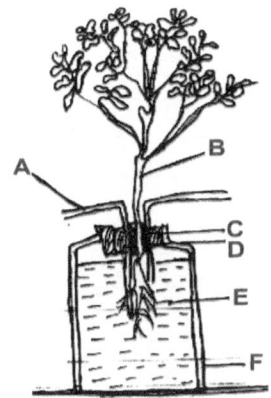

15A. What is the experimental setup above used for?

_____

15B. Identify the parts labeled A to F:

A. _____

B. _____

C. _____

D. _____

E. _____

F. _____

15C. What are the roles of A, D, and E in the experiment?

A. _____

D. _____

E. _____

15D. List three precautions that must be taken in this experiment and give a reason for each precaution:

_____

_____

_____

_____

15E. State one function for each of the elements—nitrogen, phosphorus, and calcium—in plant growth.

Nitrogen:_____

Phosphorus:_____

Calcium:_____

15F. What is meant by the terms *macro-* and *microelements?*

Macroelements:_____

_____

Microelements:_____

_____

Study the diagram below and use it to answer questions A to D.

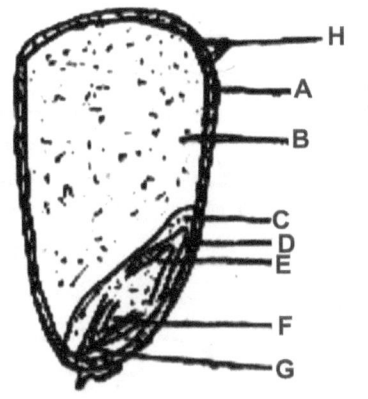

16A. Name the structures labeled A to H:

A. _____

B. _____

C. _____

D. _____

E. _____

F. _____

G. _____

H. _____

16B. To what group of flowering plants does the diagram represented above belong? _____

16C. In which of the labeled parts is food stored?

_____

16D. Outline the process of germination of the structure represented by the diagram. (No diagram is required.)

_____

_____

_____

The diagram below represents the skull of a mammal. Study it carefully and use it to answer questions A through D.

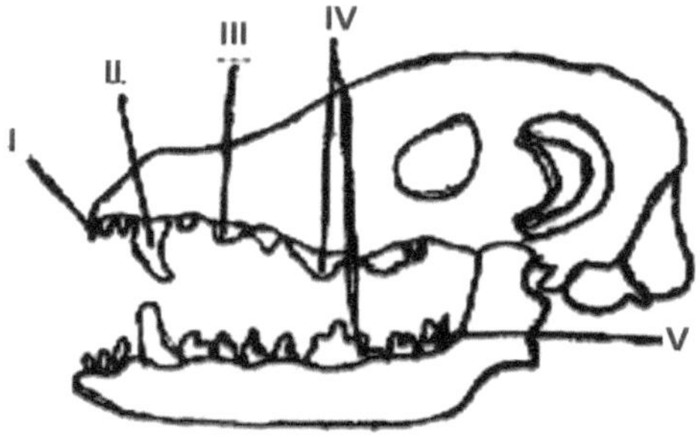

17A. Name the parts labeled I to V:

I. _____

II. _____

III. _____

IV. _____

V. _____

17B. Suggest the feeding habit of the animal:

_____

_____

17C. State the adaptations of the structures labeled I to IV to the feeding habit of the animal:

I. _____

II. _____

III. _____

IV. _____

17D. Name two animals that have the feeding habit stated above (B):

1. _____

2. _____

Study the diagram below and use it to answer questions A to D.

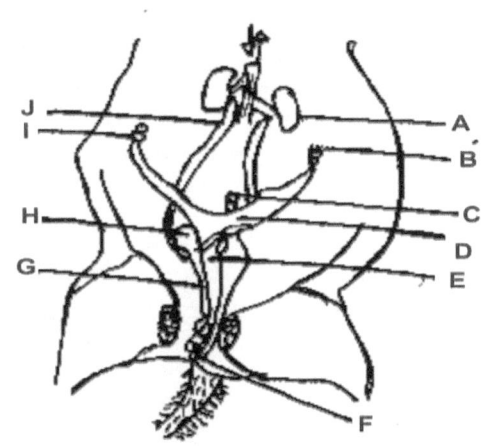

18A. Name parts labeled A to J.

A. _____

B. _____

C. _____

D. _____

E. _____

F. _____

G. _____

H. _____

I. _____

J. _____

18B. What is the sex of the animal illustrated in the diagram?

_____

18C. Name the parts of the excretory system labeled in the diagram:

_____

_____

18D. In which of the labeled parts is the gamete produced?

_____

18E. Name two of the labeled structures which are parts of the digestive system:

i. _____

ii. _____

Study the diagram below and use it to answer questions A through E.

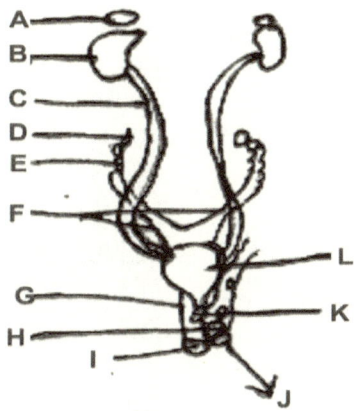

19A.

i.   Identify the structures labeled A to L:

A _____
B _____
C _____
D _____
E _____
F _____
G _____
H _____
I _____
J _____
K _____
L _____

i.   What is the sex of the animal? _____
ii. List three features that enabled you to identify the sex of the animal:

19B. State how the system illustrated above is different from that of a human being:

_____

19C. State the functions of the parts labeled E, F, G, and L:

E _____
F _____
G _____
L _____

20A. The diagram below illustrates the family pedigree of two parents, Musa and Obiageli, who have a child, Buba. Buba has sickle cell disease. The gene for normal hemoglobin is A and its recessive allele which causes sickle cell disease is S. Use the diagram to answer questions A through D.

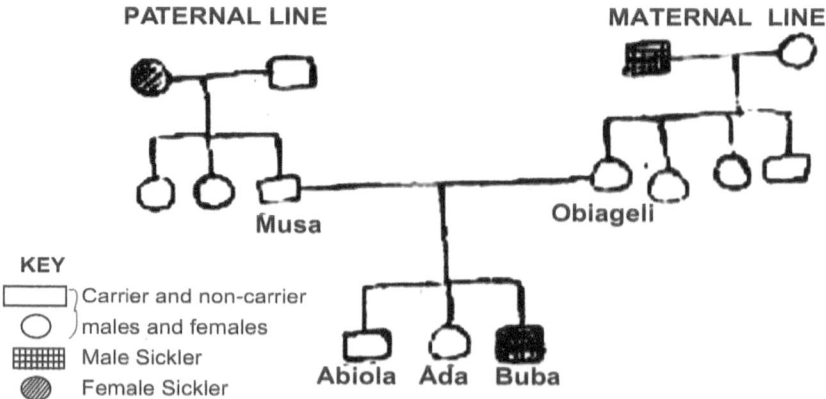

i.   With reference to sickle cell gene, state the genotypes of Musa and Obiageli:

_____

ii.   What was the genotype of Musa's father?

_____

iii.   What was the genotype of Obiageli's father?

_____

20B. Explain how it is possible for Musa and Obiageli to have a child with sickle cell disease when they do not have sickle cell disease themselves:

_____
_____
_____

20C. What are the likely genotypes of Ada and Abiola?

_____

20D. If you were a marriage counselor, what advise will you give to Ada to ensure she does not produce children with sickle cell disease?

_____
_____

Study the diagram below and use it to answer questions A to H.

21A. Name the biotic community represented in the diagram:

_____

21B. State two characteristic features of this community:

i. _____
ii. _____

21C. State two abiotic factors other than rainfall that can influence the biotic community represented in the diagram.

i. _____
ii. _____

21D. Name of instrument each for measuring the abiotic factors named in 21C.

    i.   _____

    ii.   _____

21E. Name one plant and one animal usually found in this community:

    i.   Plant:_____

    ii.   Animal:_____

21F. State one characteristic feature each of the plant and of the animal named in 21E:

    i.   Plant:_____

    ii.   Animal:_____

21G. Briefly explain how the characteristic features named above (21F) adapt the plant and animal to life in the community.

    i.   Plant:_____

    ii.   Animal:_____

Study the diagrams below and use them to answer questions A through F.

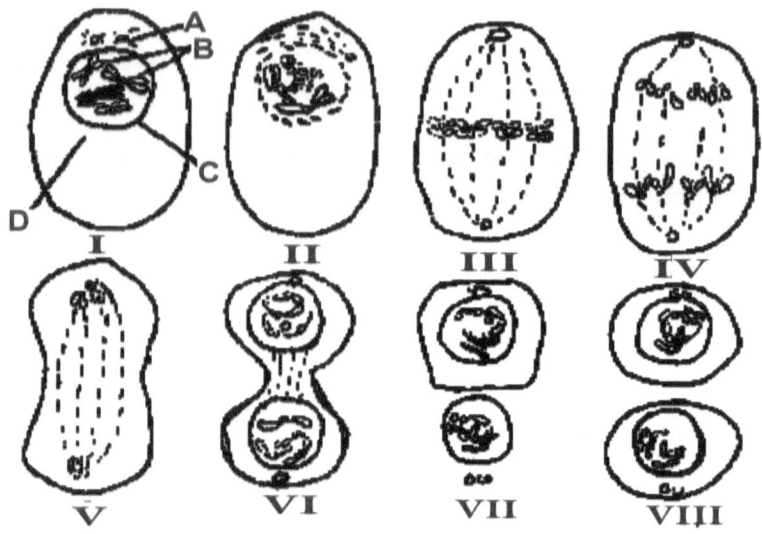

22A. What process is represented by the diagrams above?

_____

22B. State one structure each in plants and animals where this process can be observed.

Plants:_____

Animals:_____

22C. Identify the parts labeled A to D in Diagram I above:

A. _____

B. _____

C. _____

D. _____

22D. State two importance of this process in living organisms:

i. _____

ii. _____

22E. At which state illustrated in the diagram is the chromosome replicated?

_____

22F. If the chromosome number in I is 46, what will be the chromosome number each of the cells in VIII?

The diagram below is a typical scene in a farmland. Study it carefully and use it to answer questions A through E.

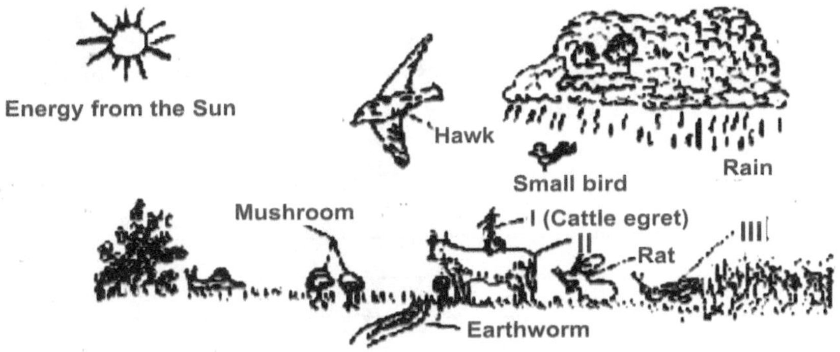

Energy from the Sun | Hawk | Small bird | Rain | Mushroom | I (Cattle egret) | Rat | Earthworm

23A. Explain why the farmland represented above could be called an ecosystem.

_____

_____

_____

23B. Explain one possible way the association between organisms I and II in the diagram could be beneficial to the two organisms.

_____

_____

_____

23C. State the term used for this association:

_____

23D. Draw a food chain using any of the organisms shown in the diagram.

23E. If there is a population explosion of the animal labeled II in this ecosystem, describe its possible effect on plants and animals in the ecosystem:

_____

_____

_____

Use the diagram below to answer question 24.

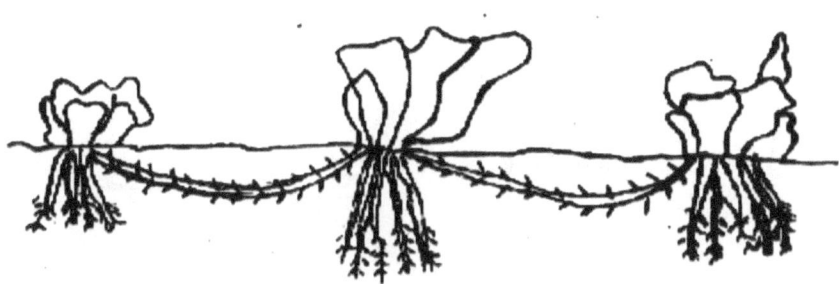

24A. Identify the organisms represented in the diagram:

_____

24B. List three features that adapt the organism to its environment.

I.   _____

II. _____

III. _____

Study the diagram below and use it to answer question 25.

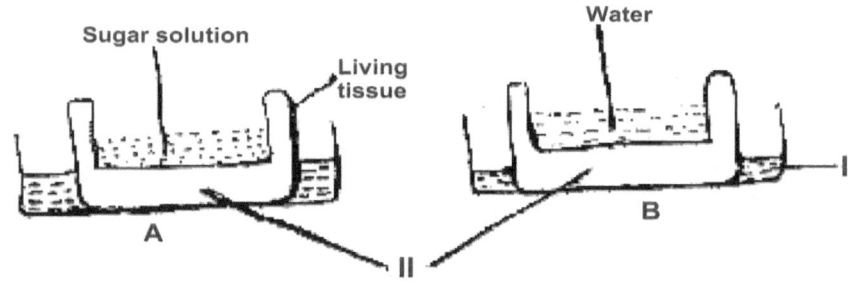

25A. What is the aim of the experiment illustrated in the diagram above?

_____

25B. Name the liquid labeled I:_____

25C. Name two materials that can be used as living tissues:

I. _____
II. _____

25D.

i.   What would happen to the level of solution inside the living tissue in both A and B if the setup is allowed to stand for about three hours?

_____
_____

ii. Give a reason for your answer:_____

25E. What is the function of setup B? _____

26A. *Study carefully the diagram below and use it to answer question 26.*

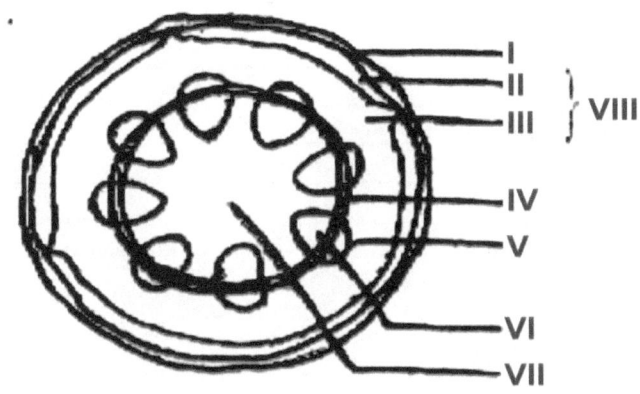

    i.   Identify the structure illustrated in the diagram above.

       _____

    ii.   Give three reasons for your answer:

       _____
       _____
       _____

26B. Name the parts labeled I to VIII in the diagram above:

I   _____   VII _____
II   _____   VIII _____
III _____
IV _____
V   _____
VI _____

26C. State one function each of the parts labeled—I, V, VI, and VIII—in the diagram above:

Function of I: _____

Function of V: _____

_____

_____

Function of VI: _____

_____

Function of VII: _____

_____

26D. What are the significance of the parts labeled II and III on the diagram?

Significance of II: _____

_____

Significance of III: _____

_____

26E.

    i.   State what happens as a result of the formation of the part labeled IV:

_____

_____

    ii. What is the importance of the process stated above to the plant?

_____

26F.

    i.   Name the reagent that could be used to identify the tissue that separates inner stele from outer stele:_____

_____

ii. What color would be observed, and what substance in the tissue is indicating this color?

_____

Study the diagram below and use it to answer the questions that follow.

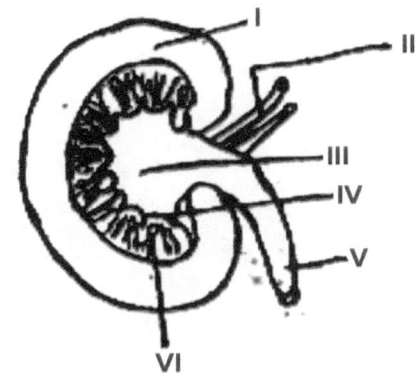

27A. Name the parts labeled I to VI:

I. _____   IV. _____
II. _____   V. _____
III. _____   VI. _____

27B. State the functions of the parts labeled II and V:

II. _____
V. _____

27C. In which organ in the mammalian body is urea formed?

_____

28A. The diagram below is a typical occurrence in nature. Study it carefully and use it to answer the questions below.

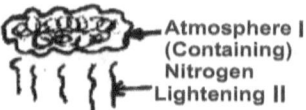

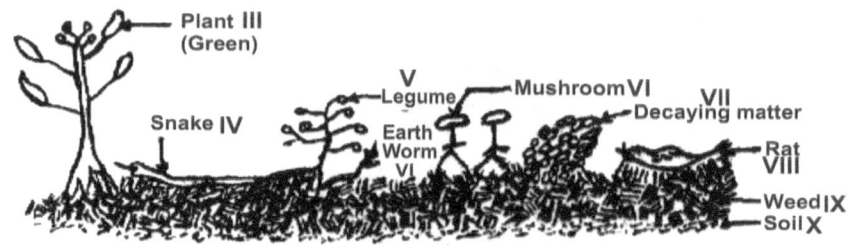

i. Name the three possible modes of nutrition carried out by the illustrated organisms in the diagram above:

I. _____

II. _____

III. _____

ii. Give three reasons for your answers above:

I   _____

II _____

III _____

28B. Construct a simple nitrogen cycle of the habitat illustrated in the diagram above using the appropriate labels in the space provided below.

28C. Explain the processes occurring in the nitrogen cycle involving the following labeled points in the diagram:

(i): I, II, and X.
(ii): IV, VIII, and X.

(i): _____

_____

_____

(ii): _____

_____

_____

28D.

    i.   Name two scavengers in the habitat:

        I.   _____

        II. _____

    ii. Draw one possible food chain in the habitat using the appropriate labels in the diagram in the space provided below.

28E.

    i.   What is the ecological name given to the diagram above apart from habitat?

_____

ii. Mention one biotic and two abiotic factors in the habitat illustrated above:

Biotic factor:_____

Abiotic factors:

I.   _____
II. _____

Study the diagram below and use it to answer questions 29A and B.

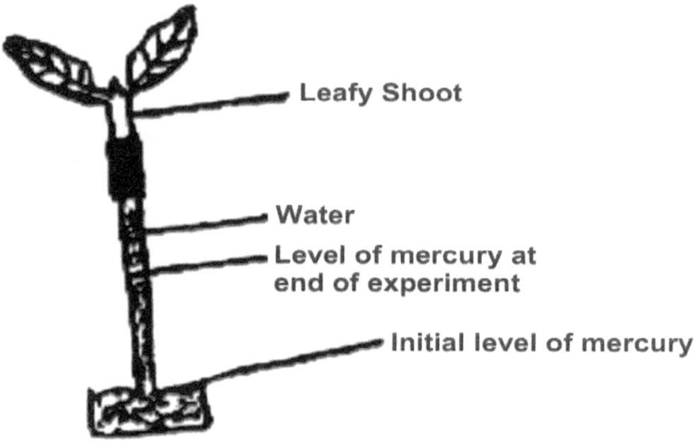

Leafy Shoot

Water

Level of mercury at
end of experiment

Initial level of mercury

29A. What is the setup used for?_____

29B. Why is the leafy shoot cut under water when setting up the experiment?

_____

29C. Name the instrument used for measuring the turbidity of water.

Diagrams I and II illustrate the digestive systems of man and bird. Use them to answer questions A to E.

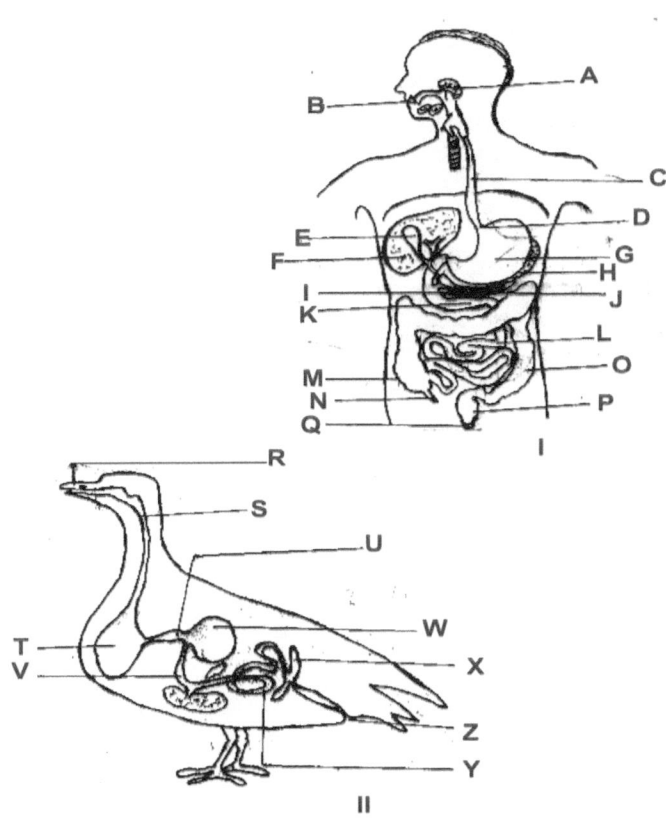

30A. Name the structures labeled A to Z:

A. _____   E. _____
B. _____   F. _____
C. _____   G. _____
D. _____   H. _____

I. _____	R. _____
J. _____	S. _____
K. _____	T. _____
L. _____	U. _____
M. _____	V. _____
N. _____	W. _____
O. _____	X. _____
P. _____	Y. _____
Q. _____	Z. _____

30B. Outline how starch and protein are digested in the digestive system of man.

Starch:

_____

_____

_____

Protein:

_____

_____

_____

30C. State the significance of the long coiled structure labeled L:

_____

_____

30D. State *two* similarities and *two* differences between the digestive system of man and that of a bird.

Two similarities:

    i. _____

    ii. _____

Two differences:

    i. _____

    ii. _____

30E.

  i. State the functions of the parts labeled N and R. (Do not limit the functions of the part labeled R to diagram II alone.)

    N _____

    R _____

  ii. What can be done if the part labeled N is inflamed?

_____

Study the information below. Use it to answer questions A to E.

In an experiment, the populations of two species of mites, A and B, were studied in the laboratory. Species A feeds on oranges, while species B feeds on species A.

Fifty individuals of species A were put in a cage and fed for fifteen days. After which, fifty individuals of species B were introduced into the same cage. The populations of both species recorded over a period of time are shown in the table below:

Time (days)	Number of Species A	Number of Species B
0	50	-
5	250	-
10	400	-
15	500	50
20	1,400	200
25	1,900	450

30	950	1,050
35	850	1,200
40	600	1,600
45	0	600
50	0	50
55	0	50
60	0	0

31A.

    i.   Why did the population of species A start to decrease after twenty-five days?

    _____

    _____

    _____

    ii. Explain why the population of species B decreased after forty days?

    _____

    _____

    _____

31B. Plot two graphs for species A and B on a single graph sheet using the data given on the same axes.

GRAPH

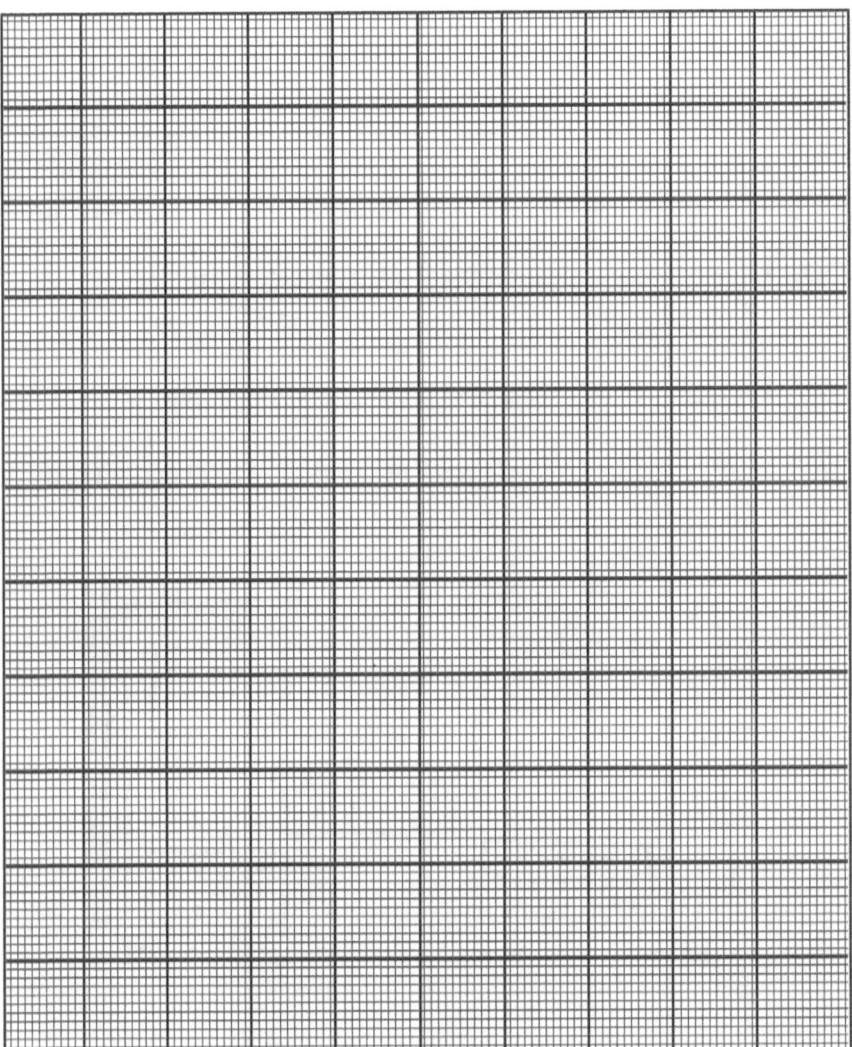

Waec Standard Graph

31C. Describe briefly the two curves you have drawn:

Curve A:_____

_____

_____

Curve B:_____

_____

_____

31D. Give one ecological term each for species A and B:

A: _____

B: _____

Study the diagram and the information below. Use them to answer question 32.

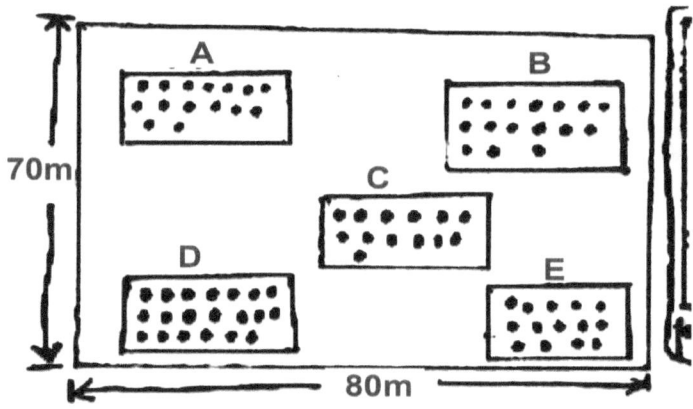

The above diagram represents an abandoned farmland of dimensions 70 m x 80 m covered by *Tridax procumbens* and other weeds.

The rectangles A, B, C, D, and E represent the areas on which a 1 x 1 m quadrat landed during a field sampling to determine the population size for *Tridax*. Only five throws of the quadrat were made during the sampling. Each dot within the rectangles A, B, C, D, and E represents a *Tridax* plant.

32A. Calculate the estimated size of *Tridax* population in the abandoned farmland.

_____

_____

_____

32B. State three reasons it is important for a farmer to know the population of the weeds on the farm:

i. _____

ii. _____

iii. _____

32C. Name two parasitic:

i. Plants:_____

ii. Animals:_____

Study the object illustrated in the diagram below and use it to answer question 33.

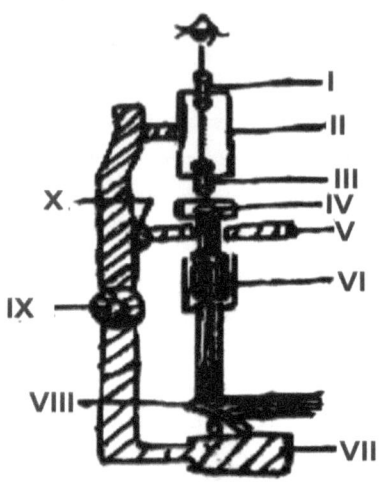

33A.

    i.   Name the object illustrated in the diagram above.

    _____

    ii.   Where is the object commonly found?

    _____

    iii. What is the object used for and why?

    _____
    _____

33B.

   i. Name the parts labeled I to X:

I.	_____	VI.	_____
II.	_____	VII.	_____
III.	_____	VIII.	_____
IV.	_____	IX.	_____
V.	_____	X.	_____

   ii. What is the function of the part labeled VII?

_____

The diagrams illustrated below represent instruments for ecological studies. Use them to answer question 34.

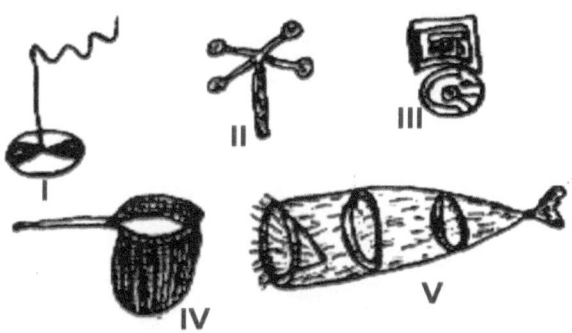

34A. Identify the instruments illustrated in diagram I to V:

I   _____

II   _____

III   _____

IV   _____

V   _____

34B. State the use of each instrument in the diagrams:

I. _____

II. _____

III. _____

IV. _____

V. _____

34C. State how the instrument illustrated in diagram I is used:

_____

_____

_____

Study the diagrams illustrated below and use them to answer question 35.

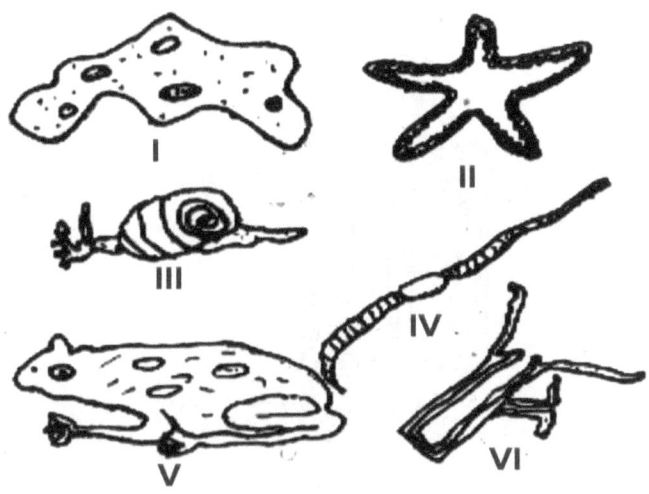

35A. Identify the organisms labeled I to VI in the diagrams illustrated, and give one reason each:

Identification	Reason
I	
II	
III	
IV	
V	
VI	

35B. Name the phylum to which each of the organisms illustrated in the diagram belongs:

I. _____   V. _____

II. _____   VI. _____

III. _____

IV. _____

Study the diagram below and use it to answer questions 36 A through D.

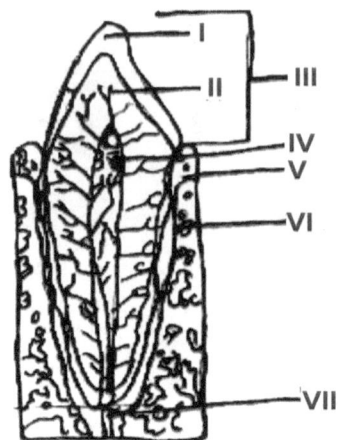

36A. Identify the structure illustrated in the diagram above:

_____

36B. Name the parts labeled I–VII:

I	_____		V	_____
II	_____		VI	_____
III	_____		VII	_____
IV	_____		VIII	_____

36C. Give one function of the structure illustrated in the diagram:_

36D.

    i.   State two functions of the part labeled IV:

_____

_____

ii.   Which of the labeled parts is least sensitive?

_____

_____

36E.

i.   Name the four types of fingerprints in man:

_____

_____

_____

ii.   State one function of fingerprints:

_____

Study the diagram below and use it to answer question 37.

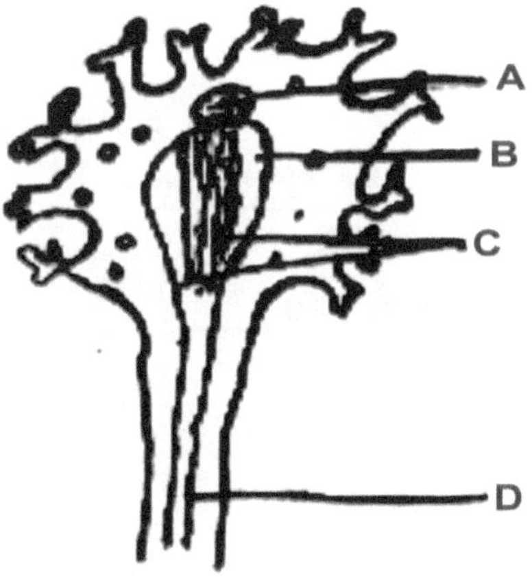

37A. Name the structure illustrated above:

_____

37B. Name the parts labeled A to D:

A. _____

B. _____

C. _____

D. _____

37C. State the function of the part labeled C:

_____

37D. In which group of animals is this structure found?

_____

Study the experiment setup illustrated below and use it to answer questions 38A, i–vi.

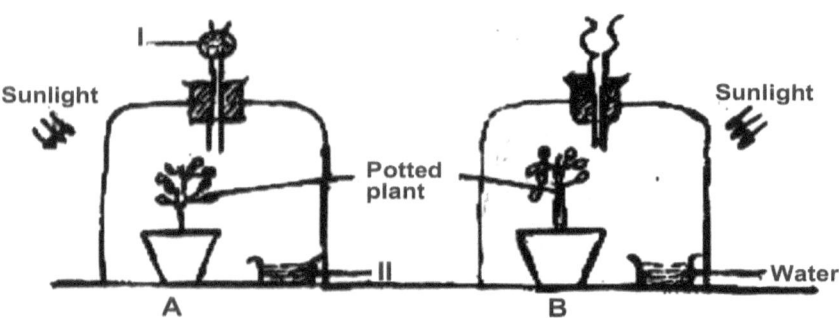

38A.

i. What is the aim of the experimental set ups in the diagram above?

_____

ii. State the substances contained in the parts labeled I and II in the setup labeled A and their functions.

Substances:

    I.   _____

    II. _____

Functions:

    I.   _____

    II. _____

iii. What purpose is the setup labeled B serving?

_____

iv. State the difference between the leaves in the setups labeled A and B at the end of the experiment:

_____

_____

Study the diagram below and use it to answer question 39A.

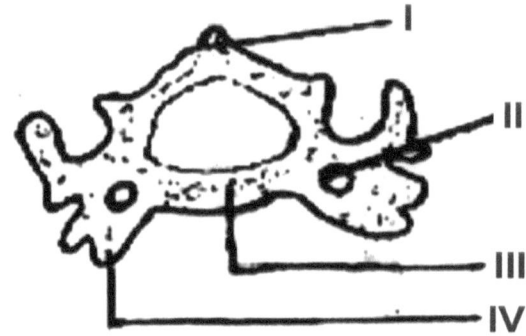

i.   Identify the structure illustrated in the diagram above:

    _____

ii. In which part of the mammalian body is it located?

    _____

iii. Name the parts labeled I–IV:

    I.   _____
    II. _____
    III. _____
    IV. _____

iv. Name the two bones which articulate with the structure illustrated in the diagram above:

    I.   _____
    II. _____

39A. State two functions of the structure illustrated in the diagram above:

_____

_____

39B. State two materials that strengthen the walls of plant cells:

I _____

II _____

39C. Name three types of phloem cells:

_____

_____

Study Diagram I below and use it to answer questions 40A, 40B, and 40C.

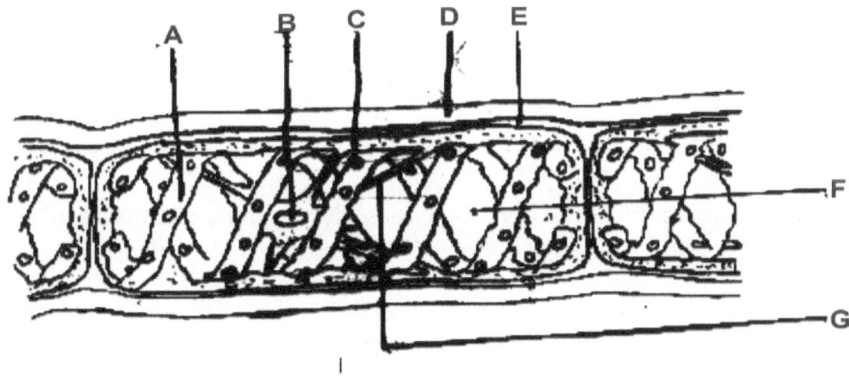

40A.

   i.   Identify the organism illustrated in the diagram (I).

_____

ii. Give two reasons for your answer in A(I) above.

_____

_____

40B.

i. Name the parts labeled A to G:

A _____

B _____

C _____

D _____

E _____

F _____

G _____

ii. Name the functions of the parts labeled A to D:

A _____

B _____

C _____

D _____

40C. To what phylum/division does the organism illustrated in diagram (I) belong?

_____

Examine diagram II.

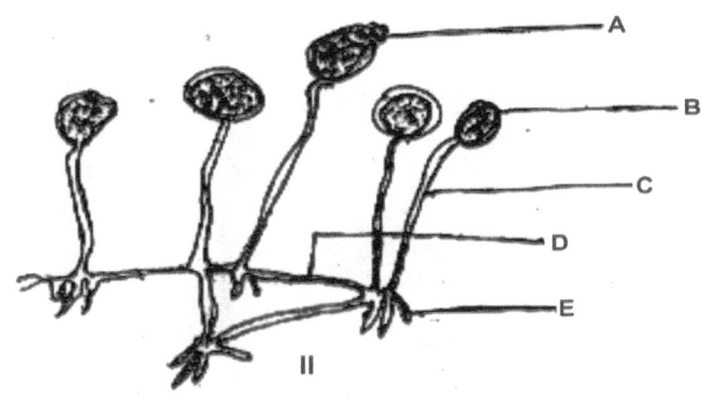

40D. Identify diagram II:_____

40E. Name the parts labeled A to E in diagram II:

A. _____
B. _____
C. _____
D. _____
E. _____

40F. Differentiate between diagrams I and II in a tabular form.

Diagram I	Diagram II

40G. Where is the organism represented in diagram II commonly found?_____

Study carefully the diagram above which illustrates an experimental setup.

41A. What is the aim of the experimental setup?

_____

41B. State two faults in the experimental setup as illustrated above:

_____

_____

41C. Name the substance labeled Z:

_____

41D. Name two chemical substances used for confirming the substance Z:

The setup in diagrams I and II illustrate breathing mechanisms in mammals.

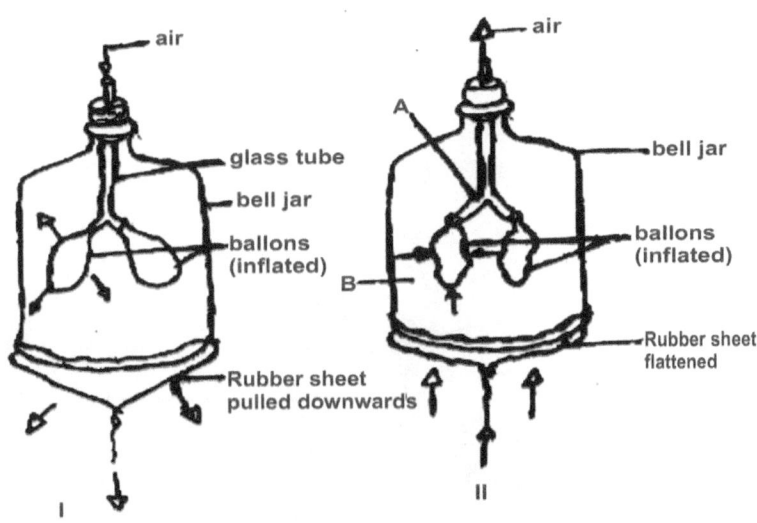

42A. Name the structures in mammals that are equivalent in function to the parts labeled:

i.   glass tube_____
ii. bell jar _____
iii. balloons _____
iv. rubber sheet_____
v.   A._____
vi. B._____

42B. Name the processes illustrated by the diagrams:

i.   I _____
ii. II _____

42C. In which of the bell jars is the pressure lower?

_____

_____

Diagram I is a drawing showing the alimentary canal of a mammal. Study the diagram carefully.

43A. Name the structures A to J:

A._____   F._____
B._____   G._____
C._____   H._____
D._____   I._____
E._____   J._____

43B. What are the functions of the parts labeled A, C, H, and J?

A. _____

C. _____

H. _____

J. _____

_____

_____

_____

Diagram II is a drawing showing the alimentary canal of a cockroach.

II

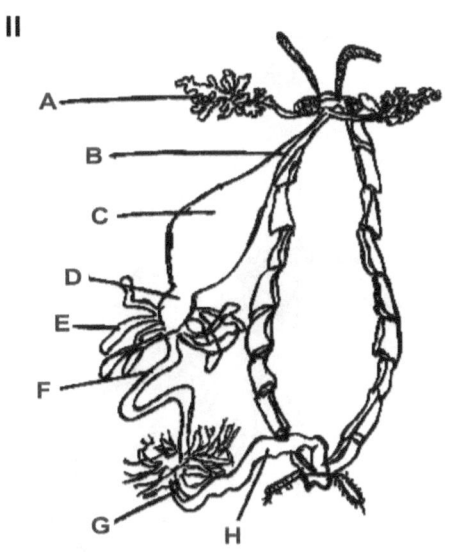

43C. Name the parts labeled A to H:

A._____   E._____

B._____   F._____

C._____   G._____

D._____   H._____

43D. State the functions of the parts labeled D and E:

D. _____

E. _____

43E. State two features common to the parts labeled C in Diagram I, and D in Diagram II.

_____

_____

43F. Identify and name the parts in Diagrams I and II that perform similar functions.

_____

Study diagram below carefully and use it to answer question 44.

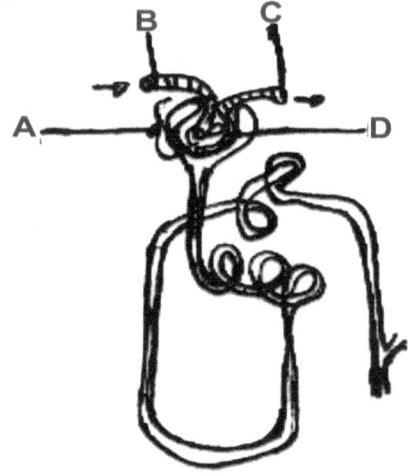

44A. Name the parts labeled A, B, C, and D:

A. _____

B. _____

C. _____

D. _____

44B. What is the significance of the relative sizes of B and C?

_____

44C. Name the structures in:

    i.   Amoeba:_____

    ii.   Earthworm:_____

Which are equivalent in function to the structure illustrated in the diagram above?

44D. Describe urine formation in man.

_____
_____
_____
_____
_____

44E.

    i.   List two other excretory organs in living organisms.

        _____
        _____

    ii.   Name four substances contained in urine:

        _____

44F. How does the kidney carry out osmoregulation in man?

_____
_____

Study the structure below carefully and use it to answer question 45.

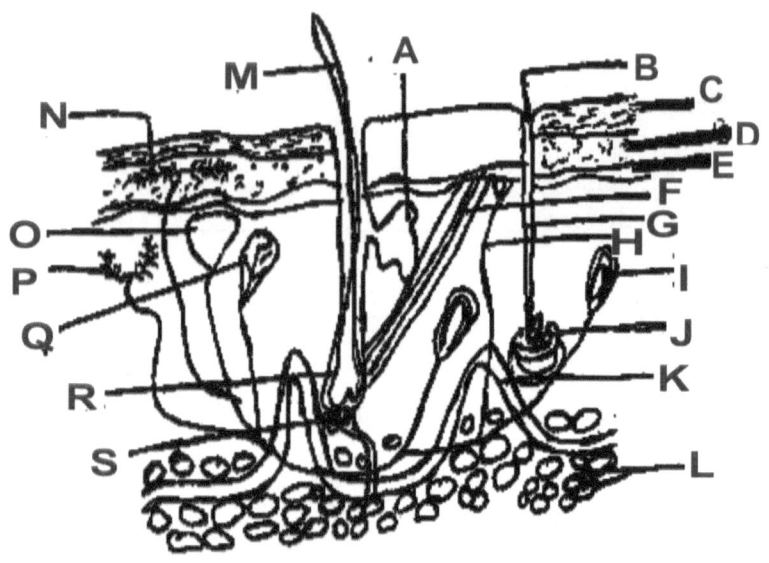

45A. Name the parts labeled A to S:

A. _____   K. _____
B. _____   L. _____
C. _____   M. _____
D. _____   N. _____
E. _____   O. _____
F. _____   P. _____
G. _____   Q. _____
H. _____   R. _____
I. _____   S. _____
J. _____

45B. State the functions of the parts labeled A, K, and L:

A _____
K _____
L _____

45C. Which parts make up the epidermis?

_____

_____

45D. Describe the structure illustrated in the diagram above.

_____

_____

_____

_____

45E. How does the structure help in the regulation of body temperature?

_____

_____

_____

_____

45F. Explain how it is possible for the structure to perform the following:

i.   Sensory function:_____

_____

ii. Excretory function:_____

_____

iii. Thermoregulation function:_____

_____

_____

_____

45G. In a tabular form, state four similarities in structure between the skin of a man and the leaf of a plant.

Skin of a man	Leaf of a plant

45H. Give five functions of the skin.

_____

_____

_____

Study the diagrams below carefully and answer questions A to E.

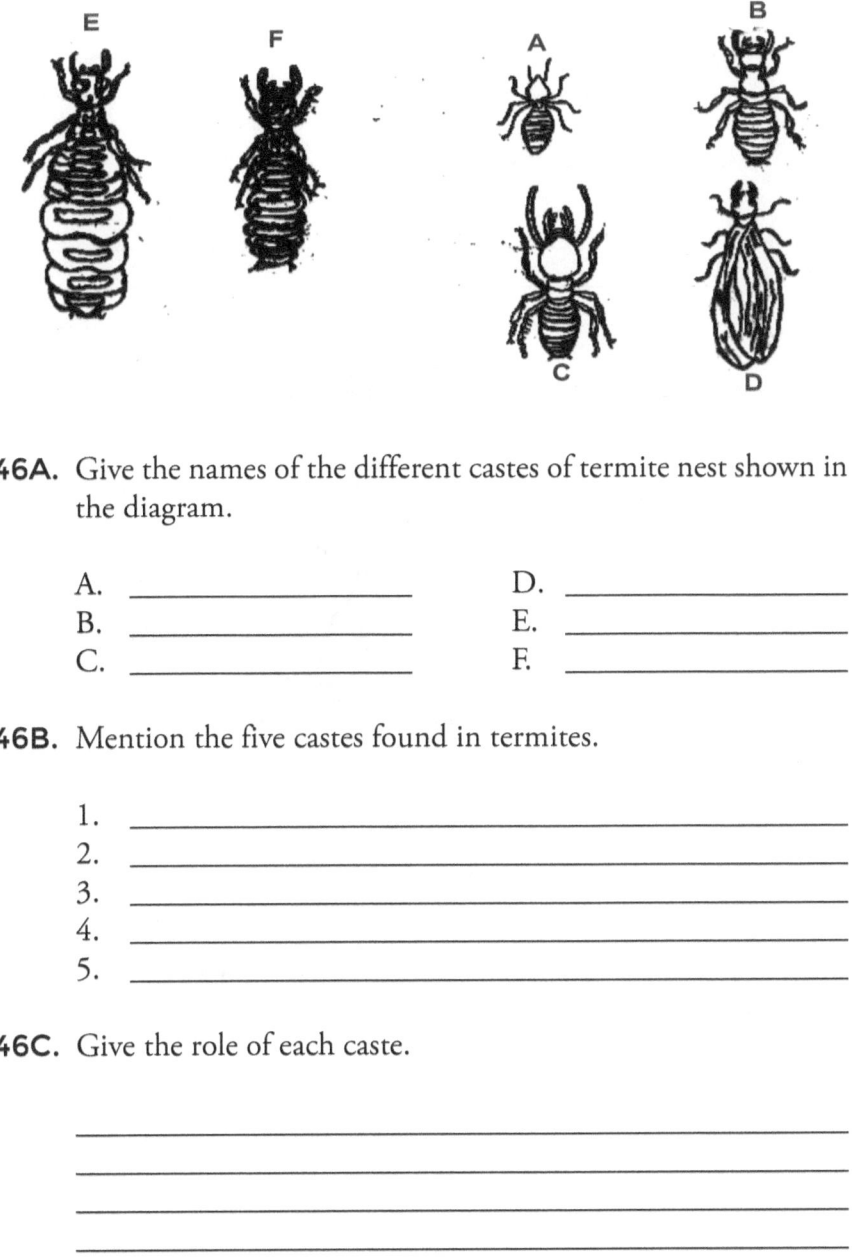

46A. Give the names of the different castes of termite nest shown in the diagram.

A. _____   D. _____
B. _____   E. _____
C. _____   F. _____

46B. Mention the five castes found in termites.

1. _____
2. _____
3. _____
4. _____
5. _____

46C. Give the role of each caste.

_____
_____
_____
_____
_____

46D. Give three economic importance of termites to man.

_____

_____

_____

46E. Why are termites regarded as social insects?

_____

_____

_____

Study the diagram below carefully and use it to answer question 47.

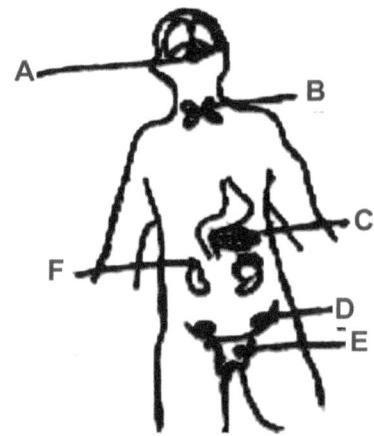

47A. Indicate the location of the following glands on the diagram above:

    A. Adrenal
    B. Pancreas
    C. Thyroid
    D. Pituitary
    E. Ovary
    F. Testes

Secretion	Gland
I. Thyroxin	
ii. Adrenaline	
iii. Insulin	
iv. Estrogen	
v. Testosterone	

47B. State one effect each of oversecretion and deficiency of:

		Oversecretion	Deficiency
i.	Thyroxin		
ii.	Adrenaline		
iii.	Insulin		
iv.	Estrogen		

47C. Give the functions of pancreas in digestion.

_____

_____

_____

Study the sketches below carefully and use them to answer questions A to E.

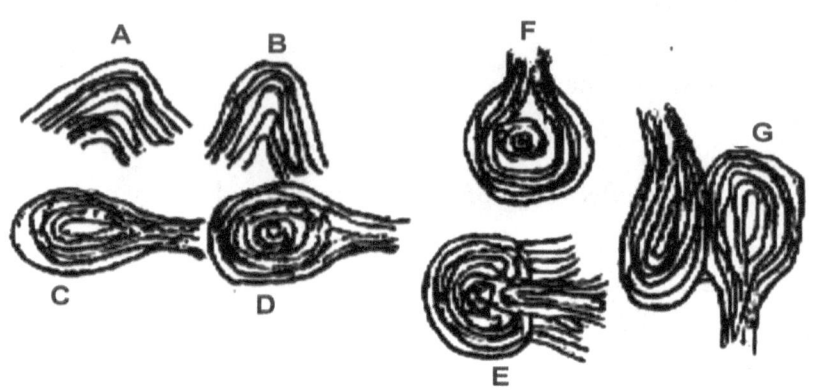

48A. What name is given to the sketches above?

_____

48B. Mention the different types based on the answer above in question A.

1. _____
2. _____
3. _____
4. _____

48C. Identify the sketches from A to G:

A _____
B _____
C _____
D _____
E _____
F _____
G _____

48D. Mention one use of the sketches to mankind and give reasons.

_____

_____

48E. What area in the study of biology are the sketches associated with?

_____

_____

The diagram below represents the transverse section of a fruit.

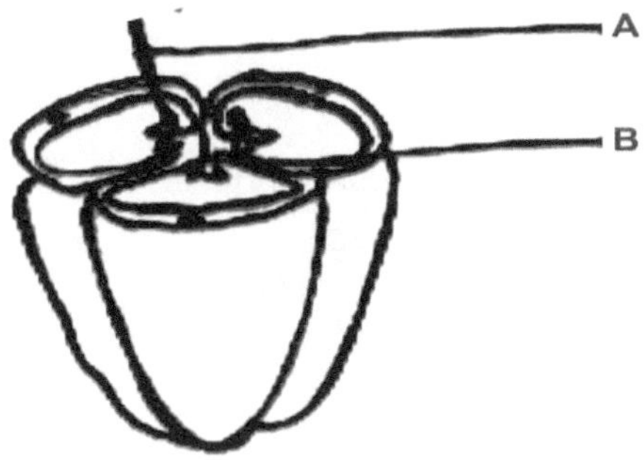

49A. State the kind of placentation found in the fruit:

_____

49B. Name one fruit that has this kind of placentation:

_____

49C. Name the structures labeled A and B:

A. _____

B. _____

49D. Define the term *placentation*:

_____

_____

49E. Mention the different types of placentation, and name one fruit each that has the kind of placentation.

Placentation	Fruits

Study the diagram below and use it to answer questions A to D.

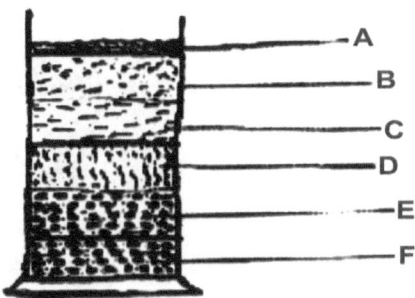

50A. Identify the parts labeled A to F.

A. _____
B. _____
C. _____
D. _____
E. _____
F. _____

50B. Suggest a suitable title for the experiment.

_____

50C. State the role(s) of the part labeled A.

_____

_____

50D. Mention four importance of soil to plants.

1. _____
2. _____
3. _____
4. _____

The diagrams below represent the transverse sections of different fruits:

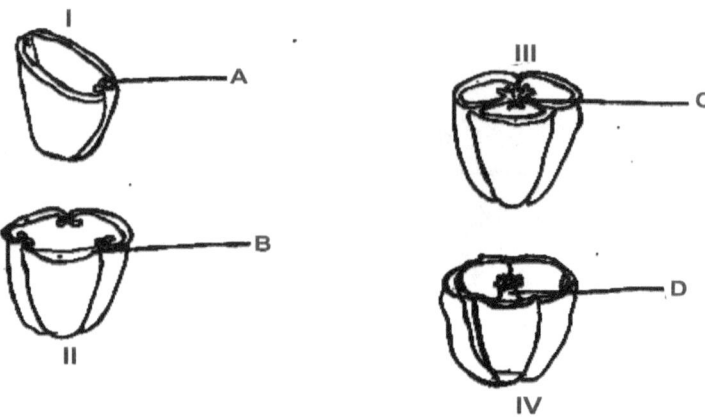

51A. Identify the parts labeled A to D:

A. _____

B. _____

C. _____

D. _____

51B. State the kinds of placentation found in the fruits:

_____

_____

_____

_____

51C. Name one fruit each that has the kinds of placentation:

_____

_____

_____

_____

Study the two diagrams below and answer questions A to K.

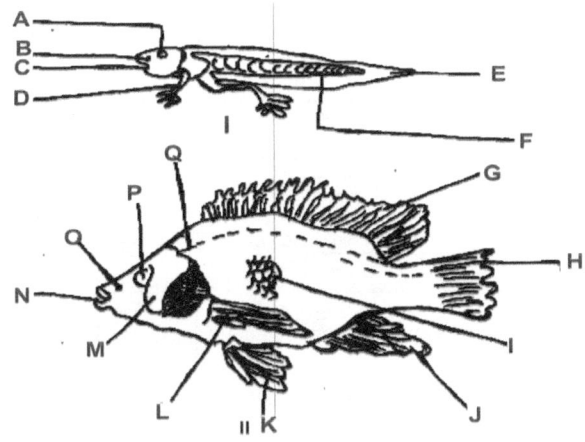

52A. Identify the organisms illustrated in the diagram:

I. _____

II. _____

52B. Name their habitats:

I. _____

II. _____

52C. Classify the illustrated organisms into their phylum and class.

I. Phylum_____ Class_____

II. Phylum_____ Class_____

52D. Give two reasons for placing Diagram II into its class based on the diagram.

1. _____

2. _____

52E. Name the parts labeled A to Q in diagrams I and II above:

A	_____	J	_____
B	_____	K	_____
C	_____	L	_____
D	_____	M	_____
E	_____	N	_____
F	_____	O	_____
G	_____	P	_____
H	_____	Q	_____
I	_____		

52F. Give one function of each of the parts labeled.

G. _____

I. _____

M. _____

52G. Name the adult stage of organism I.

_____

52H. List four external features common to organisms I and II.

1. _____

2. _____

3. _____

4. _____

52I. In a tabular form, list four observable differences between organisms I and II.

Organism I	Organism II
1.	
2.	
3.	
4.	

52J. State how four of its external features adapt organism II to its environment.

_____

_____

52K. Name two classes of food that organism II provides in human nutrition.

I. _____

II. _____

Study the diagrams illustrated below and use them to answer question 53.

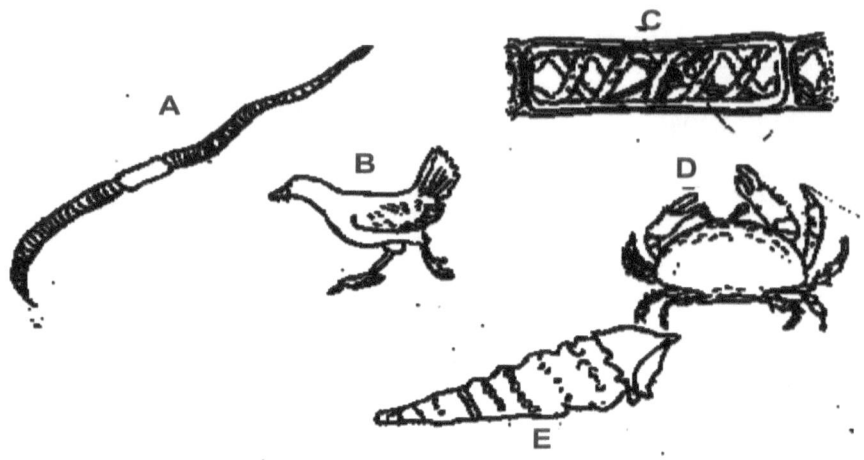

53A. Identify the organisms labeled A to E and give the phylum/ division to which each of the organism belong.

	Identification	Phylum
A		
B		
C		
D		
E		

53B.

    i.   Name the respiratory surfaces of the organisms in order of evolutionary trend.

       _____

       _____

ii. Which of the respiratory surfaces is most primitive? Give reasons.

_____

_____

53C.

i. Using specimens A to E, fill in the different organisms into their various trophic levels on the pyramid of energy constructed below using box representation.

Pyramid of Energy

ii. Name the organisms with the greatest and least amounts of the captured energy.

_____

_____

iii. List the organisms performing the following roles in the pyramid of energy above.

1. Producer:_____
2. Primary Consumer:_____
3. Secondary Consumer:_____
4. Tertiary Consumer:_____

53D. Use the second law of thermodynamics to explain the energy flow across trophic levels.

_____

_____

_____

_____

53E. Mention the three types of ecological pyramids.

1. _____

2. _____

3. _____

Study the diagram below and use it to answer question 54.

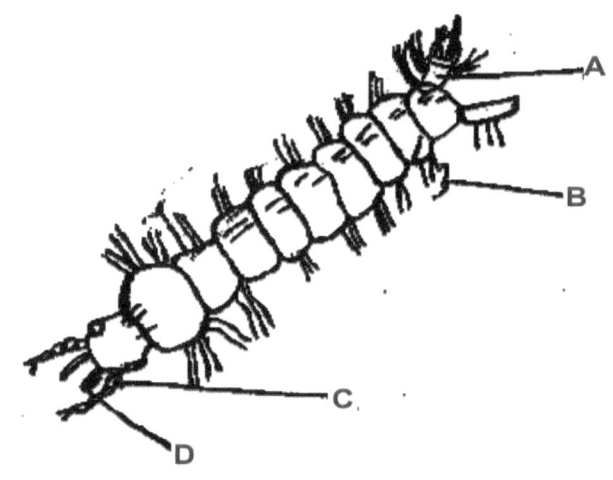

54A. Name the organism illustrated in the diagram above.

_____

54B. Name the parts labeled A to D.

A. _____

B. _____

C. _____

D. _____

54C. What is the habitat of the organism?

_____

54D. What are the functions of A and D?

A. _____

B. _____

Study the diagram below. Use it to answer questions 55A, (i–iii).

55A.

i.   Identify the organism illustrated in the diagram above.

_____

ii. What type of parasite is the organism illustrated in the diagram above?

_____

iii. Name two hosts of the organisms illustrated in the diagram above.

_____

The diagram below shows two organisms that exhibit certain association. Study the diagram carefully and use it to answer question 55 B, (i–v).

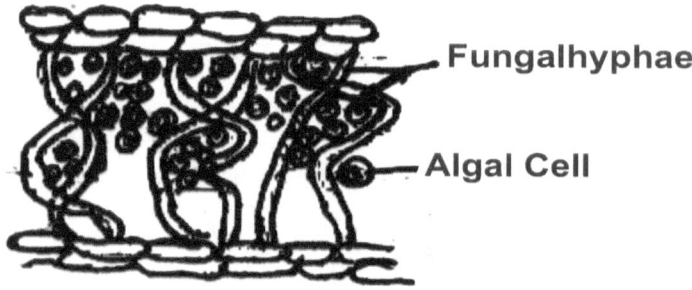

55B.

i. What type of association is shown by the organisms illustrated in the diagram above?

_____

ii. Name the organisms shown in the diagram above.

_____

iii. Name two habitats of the organism.

    1. _____

    2. _____

iv. Name two plant organisms and one animal organism in which this type of association is exhibited.

Plant organisms.

    I. _____

    II. _____

Animal organism.

_____

v. Name two other forms of associations among organisms apart from the one mentioned above.

    1. _____

    2. _____

Study the different forms of fungi below and use them to answer question A to H.

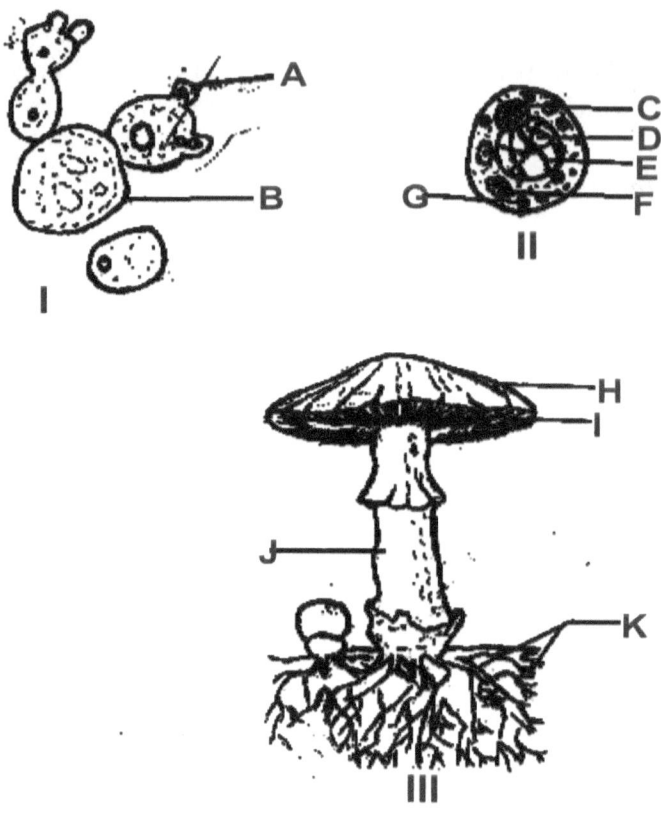

56A. Name the organisms illustrated in the diagram above.

    i. _____

    ii. _____

    iii. _____

56B. To what phylum/division do they belong?_____

56C. Name the parts labeled A to K.

A. _____   G. _____
B. _____   H. _____
C. _____   I. _____
D. _____   J. _____
E. _____   K. _____
F. _____

56D. What are the characteristics of the group?

_____

_____

_____

_____

56E. Where can diagrams I and II be naturally found?

_____

_____

56F. Give two importance of Diagrams I and II.

1. _____
2. _____

56G.

i.   Name the habitat of Diagram III.

_____

ii.   Described the structure of Diagram III.

_____
_____
_____
_____

56H. Give other examples of organisms in the group fungi.

_____
_____

Study carefully the diagram below and use it to answer question 57A.

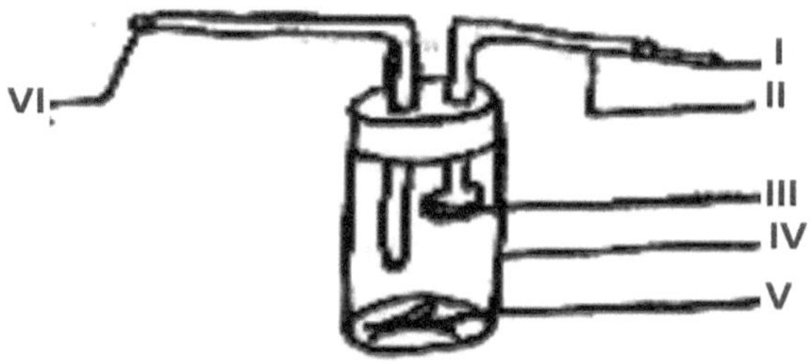

57A.

i.   Identify the apparatus illustrated in the diagram above.

_____

ii.   Name the parts labeled I–VI in the diagram.

I.   _____
II.   _____
III.   _____
IV.   _____
V.   _____
VI.   _____

iii. What is the apparatus used for?

_____

The diagram below represents an instrument used in an ecological study. Use it to answer questions A to D.

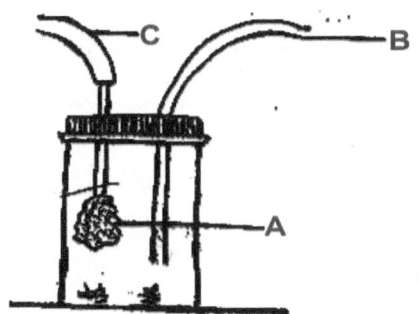

58A. What is the name of the instrument?

_____

58B. What is the instrument used for?

_____

58C. What is the function of the part labeled A?

_____

58D. Name the parts labeled B and C.

B. _____

C. _____

Study carefully the diagram below. Use it to answer question 59.

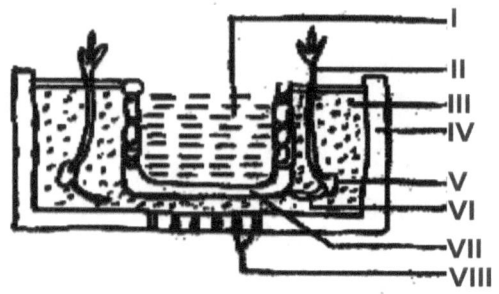

59A.

    i.   Identify the experimental setup illustrated in the diagram above.

    _____

    ii. Name the parts labeled I to VIII.

I.	_____	V.	_____
II.	_____	VI.	_____
III.	_____	VII.	_____
IV.	_____	VIII.	_____

59B. State two precautions taken to ensure accurate results when performing such experiment.

_____

_____

_____

_____

59C.

    i.   What will happen to the part labeled VI if the structure labeled I is absent?

_____

_____

    ii.   Which of the labeled parts, I or II, is more important?___

59D. State the aim of the experimental setup.

_____

_____

A food substance was put in two test tubes labeled I and II. In test tube I, Fehling's solutions, A and B, were added, and the mixture warmed. A negative result was obtained.

In test tube II, dilute hydrochloric acid was added, and the mixture was boiled. After which, few drops of sodium hydroxide were added. Then Fehling's solutions A and B were added. A positive result was obtained in test tube II after warming.

60A. What color should indicate a positive result in test tube II?

_____

60B. What food substance must have given the positive result in test tube II?

_____

60C. What role did dilute Hydrochloric acid play in the test?

_____

60D. Why was sodium hydroxide added to test tube II?

_____

60E. Why was a negative result obtained in test tube I?

_____

60F. Suggest the food substance in the original samples tested.

_____

Below is a list of laboratory reagents: dilute sodium hydroxide, Fehling's solution, Millon's reagent, iodine solution, Benedict's solution, copper sulphate solution, Sudan III Solution, dilute hydrochloric acid.

Write down the reagent(s) used in carrying out one test for each of the following:

a. Starch:_____

b. Reducing sugar:_____

c. Protein:_____

d. Fat:_____

e. Complex sugar:_____

Study the diagram of a rat's brain below and answer questions A and B.

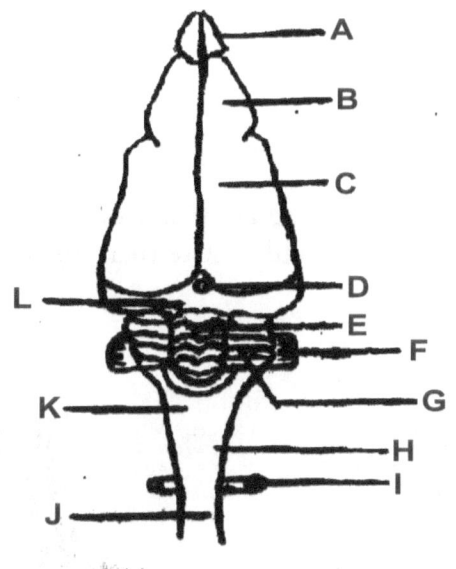

61A. Name the parts labeled A to L.

A_____   G_____
B_____   H_____
C_____   I_____
D_____   J_____
E_____   K_____
F_____   L_____

61B. Give two functions each of:

i.   Thalamus:_____

ii. Medulla oblongata:_____

iii. Hypothalamus:_____

iv. Cerebellum in man:_____

61C. Briefly describe the mechanism of transmission of impulses by neurons.

_____

_____

_____

_____

The diagrams below show the anterior views of different bones of the mammalian vertebral column. Use them to answer questions A through C.

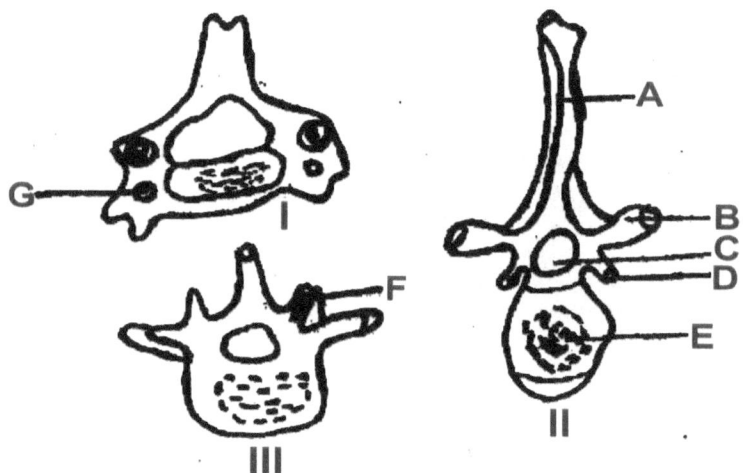

62A. Identify the bones labeled I, II, and III; and in each case, state one distinguishing characteristic.

i. _____

ii. _____

iii. _____

62B. Identify the structures labeled A to G in the Diagrams I, II, and III and state their functions.

A. _____

B. _____

C. _____

D. _____

E. _____

F. _____

G. _____

62C. Give three differences between I and II.

Differences:

I	II

Study carefully the structure of a neuron below and answer questions A to D.

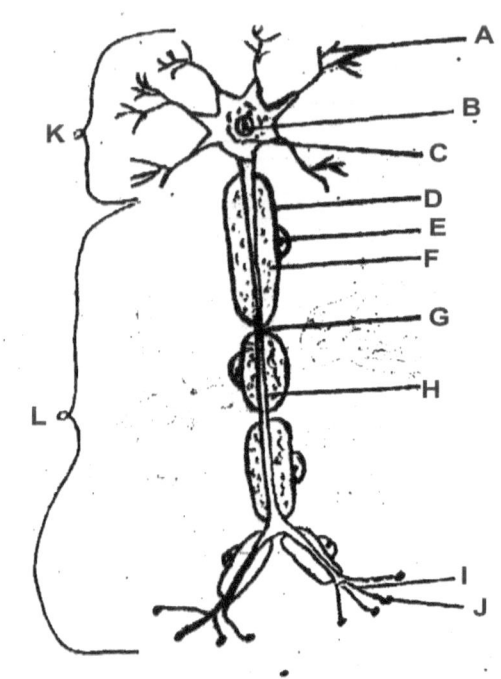

63A. Name the parts labeled A to L.

A. _____   G. _____
B. _____   H. _____
C. _____   I. _____
D. _____   J. _____
E. _____   K. _____
F. _____   L. _____

63B. Describe the structure of a neuron._____
_____
_____

63C. Name the three types of neurons and give the function of each.

_____

_____

_____

_____

Identify the bones labeled A and B, and in each case, give four observable features each of A and B.

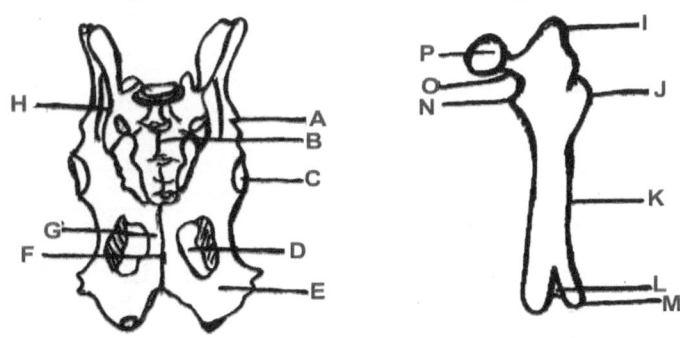

64A.

A. _____

Observable features.

1. _____
2. _____
3. _____
4. _____

B. _____

Observable features.

1. _____
2. _____
3. _____
4. _____

64B. Identify the parts labeled A to P.

A. _____   I. _____
B. _____   J. _____
C. _____   K. _____
D. _____   L. _____
E. _____   M. _____
F. _____   N. _____
G. _____   O. _____
H. _____   P. _____

64C. State three functions each of the structures labeled A and B.

Bone A

1. _____
2. _____
3. _____

Bone B

1. _____
2. _____
3. _____

64D. State how the structure labeled A is adapted to articulate with the structure labeled B.

_____

_____

_____

64E. Name a nutrient contained in bones A and B.

What function does this nutrient perform in the human body?

Nutrient

A. _____

B. _____

Function

_____

_____

_____

Study the diagrams below and use them to answer questions A to D.

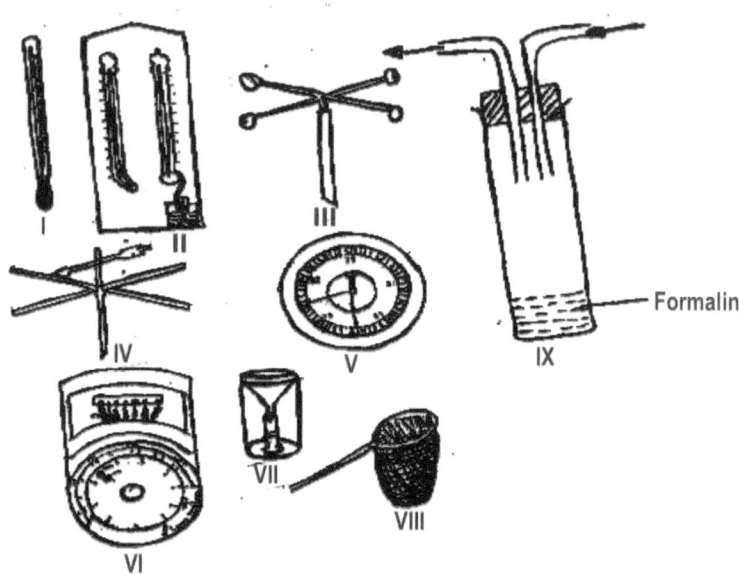

Formalin

65A. Identify the instruments represented as I to VIII:

I. _____

II. _____

III. _____

IV. _____

V. _____

VI. _____

VII._____

VIII. _____

65B. What are the uses of the instruments represented as I to VIII?

I. _____

II. _____

III. _____

IV. _____

V. _____

VI. _____

VII._____

VIII. _____

65C. Describe how I and VII could be used in a field trip for measuring abiotic factor.

*Instrument I:*_____

_____

_____

_____

*Instrument VII:*_____

_____

_____

_____

65D. Diagram IX is supposed to be a pooter. List three faults that can be observed from the diagram of the setup.

_____

_____

_____

Study the diagrams below and use them to answer questions A through F.

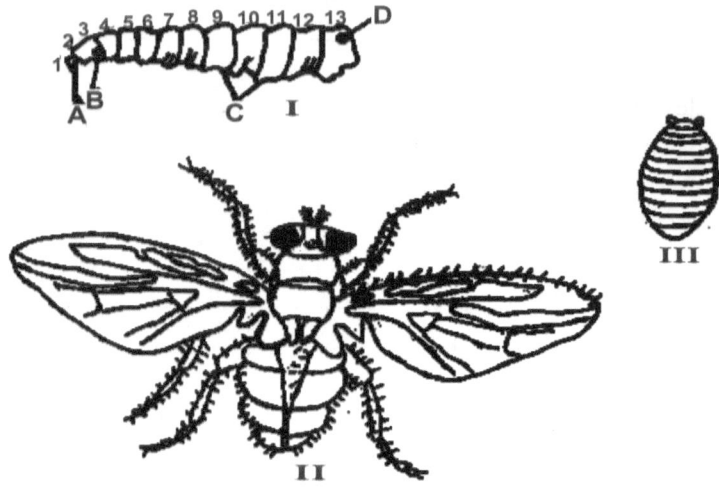

66A.

    i.   Identify the organisms shown in the diagrams.

          I.   _____

          II. _____

          III. _____

    ii.   Identify the structures labelled A, B, C, and D.

          A. _____

          B. _____

          C. _____

          D. _____

66B. What is the function of the structures labeled A, B, C, and D?

A. _____

B. _____

C. _____

66C. State one harmful economic importance of the organism labeled in the diagram as II.

_____

66D. State one adaptive features of organism II visible in the diagram which helps the organism carry out its harmful economic importance.

_____

_____

_____

66E. Explain how this harmful effect is carried out by this organism with the help of the adaptive feature identified above.

_____

66F. What is the color of I and II in live specimens?

I. _____

II. _____

The setup below was used to demonstrate that carbon oxide (iv) is given during germination of seeds. The setup was left to stand for twelve hours.

Use the drawings to answer questions A to E.

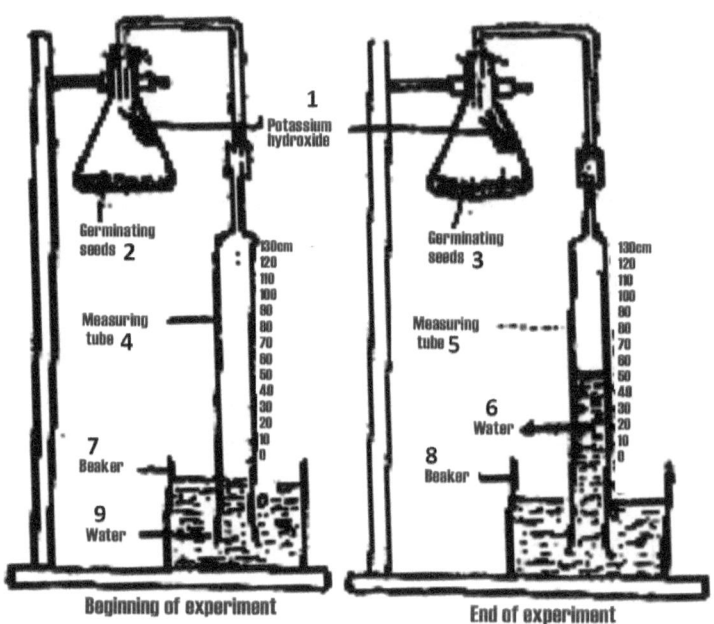

Beginning of experiment   End of experiment

67A. State one precaution which was taken to ensure the success of the experiment.

67B. What is the volume of carbon oxide (iv) produced in this experiment?

67C. Calculate the rate of production of carbon oxide (iv) per hour assuming the gas was produced at a constant rate during the experiment.

_____

_____

_____

67D. Explain in detail how this setup works in determining the volume of carbon oxide produced in this experiment.

_____

_____

_____

67E. List three conditions necessary for germination of seeds.

i. _____

ii. _____

iii. _____

The diagram below is the profile of a freshwater habitat. Use it to answer questions A through G.

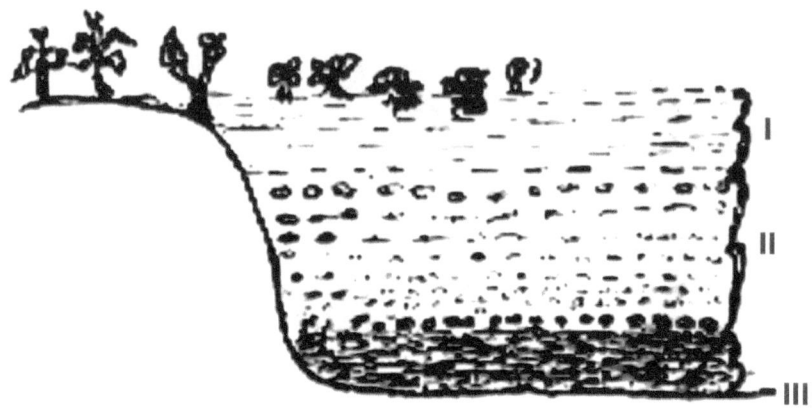

68A. Identify the zones labeled I and II.

I. _____

II. _____

68B. List two abiotic characteristics of Zone I.

_____

_____

68C. Mention two plants and two animals usually present in Zone I.

*Plants:*_____

_____

*Animals:*_____

_____

68D. Which of the zones has the highest level of primary production?

_____

68E. Explain your answer above.

_____

_____

68F. State two differences between plants in Zone I and plants in the tropical rainforest.

Plant in Zone I	Plants in Tropical Rainforest

68G. Name three organisms usually found in Zone III.

1. _____
2. _____
3. _____

Study the diagrams below carefully. Use them to answer questions A to D.

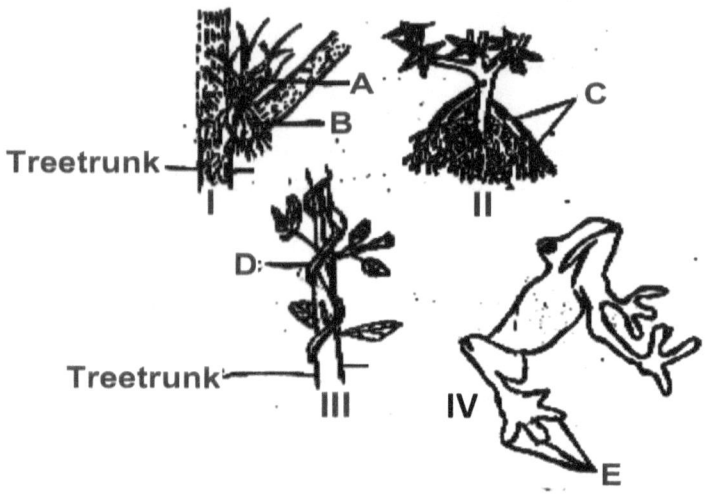

69A.

    i.   Identify the organisms represented as I to IV.

          I.   _____

          II. _____

          III. _____

          IV. _____

    ii. Name the parts labeled A, B, C, and E.

          A. _____

          B. _____

          C. _____

          E. _____

69B. Name the biotic community in which each organism labeled I to IV is usually found.

I. _____

II. _____

III. _____

IV. _____

69C. State three characteristics of the biotic community in which the organism labeled III is usually found.

I. _____

II. _____

III. _____

69D. Name the special adaptive features of the parts labeled B to E, and explain how each feature adapts the organism to its habitat.

B. _____
_____

C. _____
_____

D. _____
_____

E. _____
_____

Study the two diagrams below and answer questions A to I.

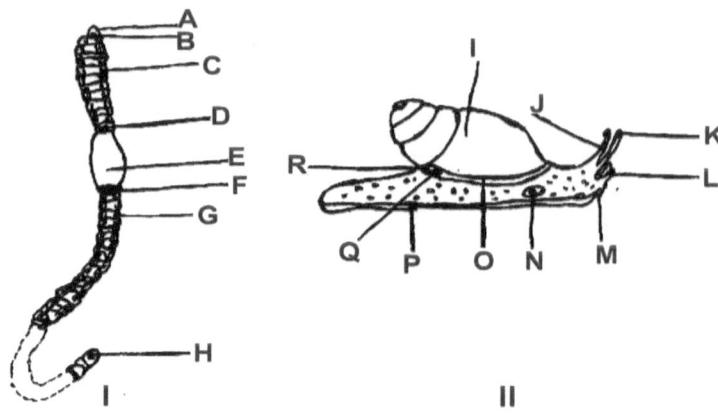

70A. Identify the diagrams above.

I. _____

II. _____

70B. Name the phylum to which the organisms above belong.

I. _____

II. _____

70C. Identify the parts labeled A to R.

A. _____   J. _____
B. _____   K. _____
C. _____   L. _____
D. _____   M. _____
E. _____   N. _____
F. _____   O. _____
G. _____   P. _____
H. _____   Q. _____
I. _____   R. _____

70D. State the functions of the parts labeled C and E.

C _____

E _____
_____

70E. Name the habitats of the two organisms.

I. _____
II. _____

70F. Which alphabets indicate the male and the female sex organs in the diagram labeled I? What is the term used to describe such an organism?

_____
_____

70G. Why is the organism in Diagram I referred to as nocturnal animal?

_____

State two external features of the diagram labeled II which are protective in function.

_____
_____

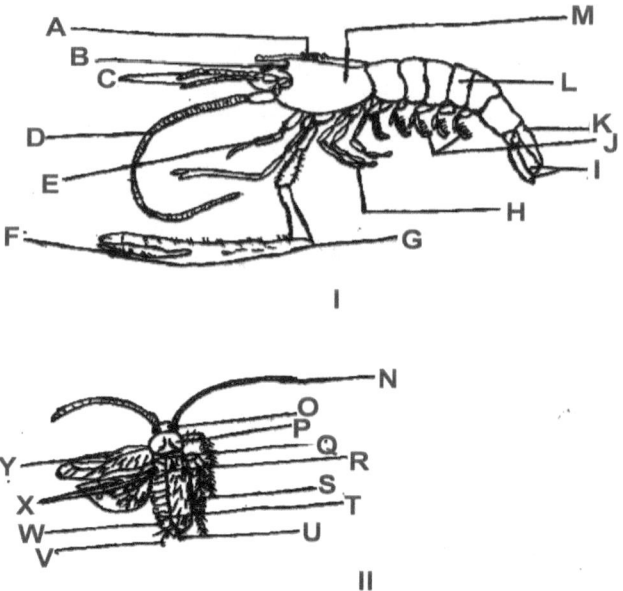

71A. Identify the two organisms above.

I. _____

II. _____

71B. Name the habitats of the two organisms.

I. _____

II. _____

71C. Identify the parts labeled A to Y.

A. _____   H. _____

B. _____   I. _____

C. _____   J. _____

D. _____   K. _____

E. _____   L. _____

F. _____   M. _____

G. _____   N. _____

O. _____   U. _____
P. _____   V. _____
Q. _____   W. _____
R. _____   X. _____
S. _____   Y. _____
T. _____

71D. Name the phylum and class to which the organisms belong.

	Phylum	Class
I		
II		

71E. Give the economic importance of the class that the organism labeled II belong.

_____

_____

_____

71F. Mention the body divisions in the two diagrams.

I   _____

II _____

71G. Mention the different mouthparts in the organism labeled II.

_____

71H. Mention four examples each of organisms that belong to the same class with the two organisms.

_____

_____

71I. List three external features common to diagrams labeled I and II which show that they are arthropods.

_____

_____

_____

71J. What is the sex of the organism labeled II? Give reason for your answer.

_____

_____

71K. Name the special adaptive features of the organisms labeled II and explain how each feature adapts the organism to its habitat.

_____

_____

_____

_____

Study the diagram below and use it to answer questions A to E.

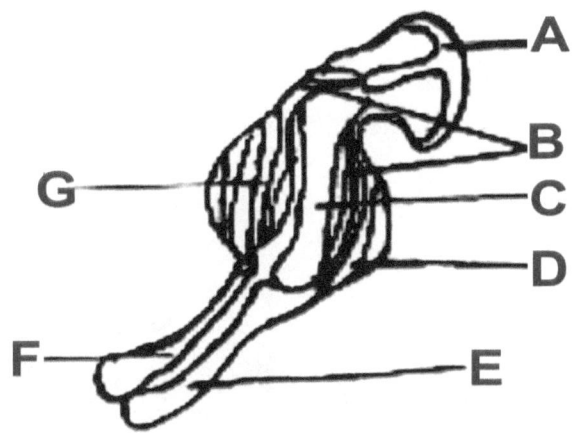

72A. Name the structures labeled A to G in the diagram above.

A. _____   E. _____
B. _____   F. _____
C. _____   G. _____
D. _____

72B. Name the type of joint between A and C.

_____

72C. Mention two other types of joints in the mammalian body and their locations.

_____

_____

72D.

    i.   State the name of the fluid found in the joint.

    _____

    ii. What is the function of the joint?_____

Study the diagram below and answer questions A to E.

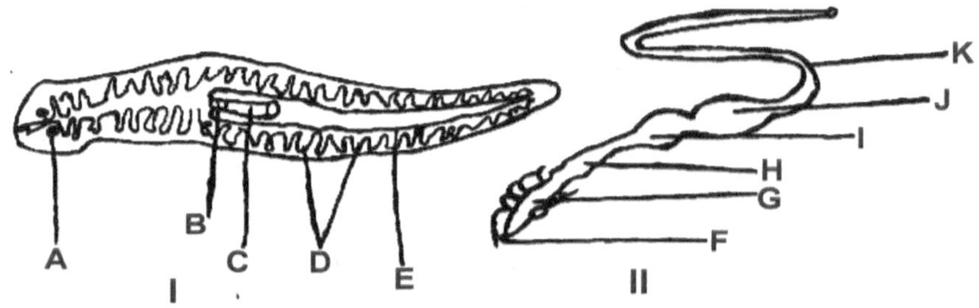

73A. Identify the diagrams illustrated above.

    I   _____

    II   _____

73B. Identify the parts labeled A to K.

    A. _____   K. _____

    B. _____

    C. _____

    D. _____

    E. _____

    F. _____

    G. _____

    H. _____

    I. _____

    J. _____

73C. How many branches does the intestine of the organism labeled I have, and what does the organism feed on?

_____

73D. State the functions of the parts labeled I and J.

    I.   _____

    J.   _____

73E. In which of the organism does diffusion of digested food take place?

_____

Study the diagrams illustrated below and use them to answer question 74.

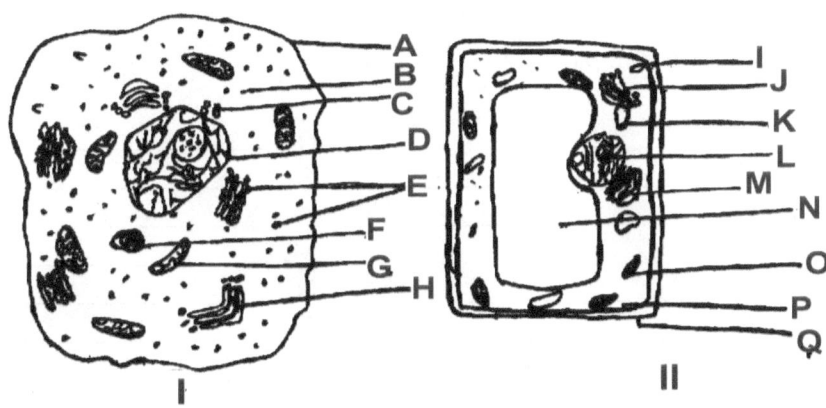

74A. Identify the diagrams illustrated above.

    I.   _____

    II.   _____

74B. List out the similarities and the differences between the structures illustrated above.

Similarities

_____

_____

_____

Differences

I	II

74C. Identify the parts labeled A to Q.

A. _____   J. _____

B. _____   K. _____

C. _____   L. _____

D. _____   M. _____

E. _____   N. _____

F. _____   O. _____

G. _____   P. _____

H. _____   Q. _____

I. _____

74D. What does the cell theory say?

_____

_____

_____

74E. Mention five scientists that have contributed to the study of cells.

1. _____

2. _____

3. _____

4. _____

5. _____

74F. Give the functions of the following parts.

A. _____
   _____

C. _____
   _____

D. _____
   _____

E. _____
   _____

F. _____
   _____

J. _____
   _____

K. _____
   _____

M. _____
   _____

N. _____
   _____

O. _____
   _____

75A.

    i.   Name the different blood cells in humans.

_____

_____

    ii. What is the fluid portion of blood called?

_____

    iii. Which blood groups are universal donors and universal recipients?

_____

    iv. Give reasons for your answers to 75A, iii above.

_____

_____

75B. What are the functions of the mammalian blood?

_____

_____

_____

_____

75C. Complete the table below to show the blood group compatibility. Use a tick (✓) compatibility and the letter X for incompatibility.

Recipient Blood Group	Indicate Antibodies Present in the Spaces	Donor's Blood Group			
		A		AB	O
A					
B					
AB					
O					

Study the diagram illustrated below carefully and answer questions A to G.

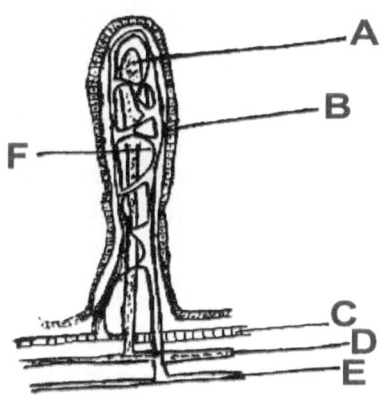

76A. Identify the structure above.

_____

76B. Where is the structure located in the mammalian body?

_____

76C. List the enzymes that are produced in your answer above and the substrates they act on:

_____

_____

76D. List the characteristics of enzymes:

_____

_____

_____

76E. Identify the parts labeled A to F.

A. _____   D. _____

B. _____   E. _____

C. _____   F. _____

76F. State the adaptive features possessed by the small intestine that enables it to perform the function of absorption.

_____

_____

_____

76G. Mention the products of digestion that are absorbed in A, B, and F.

A. _____

B. _____

F. _____

Study the diagrams illustrated below and use them to answer question 77.

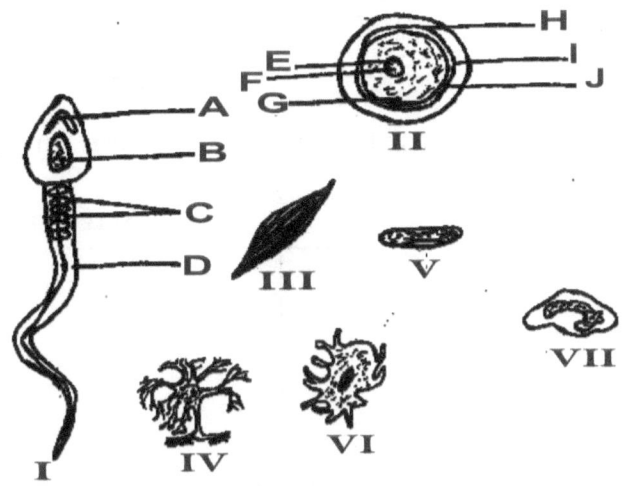

77A. Using the diagrams illustrated above, fill in the table below.

Cell	Shape	Location	Function
I.			
II.			
III.			
IV.			
V.			
VI.			
VII.			

77B. Identify the parts labeled A to J.

A. _____   G. _____
B. _____   H. _____
C. _____   I. _____
D. _____   J. _____
E. _____
F. _____

77C. Give the functions of the parts labeled A, C, and D.

A._____

B._____

D._____

77D. In a tabular form, give the differences between I and III.

I	II

Study the diagram below and use it to answer question 78.

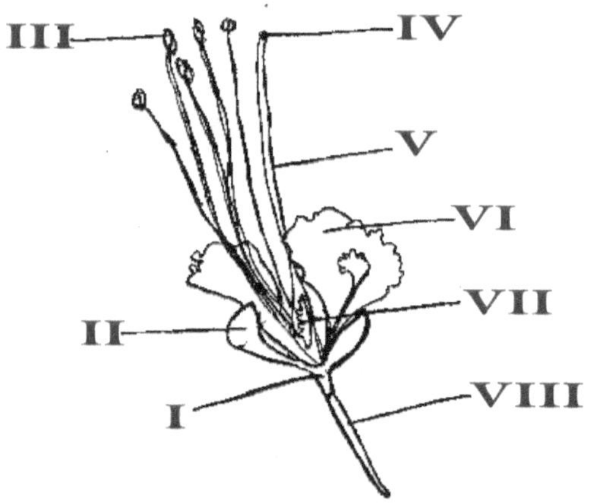

78A. Identify the parts labeled I to VIII.

I. _____   V. _____

II. _____   VI. _____

III. _____   VII._____

IV. _____   VIII. _____

78B. Which of the labeled parts represent the:

I.   female reproductive system:_____

II. petal:_____

78C. Mention the four floral parts of a flower.

_____

78D. Give five differences between wind- and insect-pollinated flowers.

Wind-Pollinated Flower	Insect-Pollinated Flower

78E. Give two examples each of wind- and insect-pollinated flowers.

_____

_____

_____

78F. Define the term *flower*._____

78G. Give the functions of the parts labeled I to VI.

I.   _____

II.   _____

III. _____

IV.   _____

V.   _____

VI.   _____

Use the diagram above to answer questions A to G.

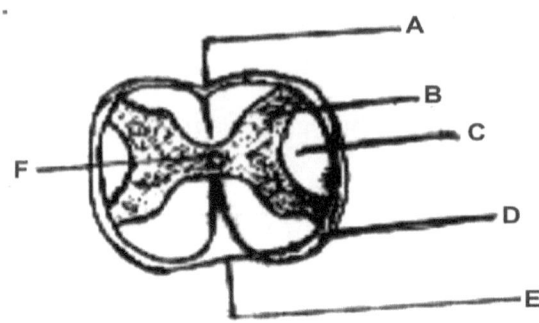

79A. Identify the structure illustrated above.

_____

79B. List any three differences between a reflex action and a voluntary action:

Reflex Action	Voluntary Action

79C. Give the differences between somatic and the autonomic systems.

Somatic System—Autonomic System

_____

_____

_____

79D. What are the functions of sympathetic and the parasympathetic nervous systems?

_____
_____
_____
_____

79E. Give two functions of the spinal cord.

_____
_____

79F. Give five examples each of reflex actions and voluntary actions.

_____
_____
_____
_____

79G. Identify the structures labeled A to F.

A. _____
B. _____
C. _____
D. _____
E. _____
F. _____

Carefully study the diagram below and answer the questions that follows.

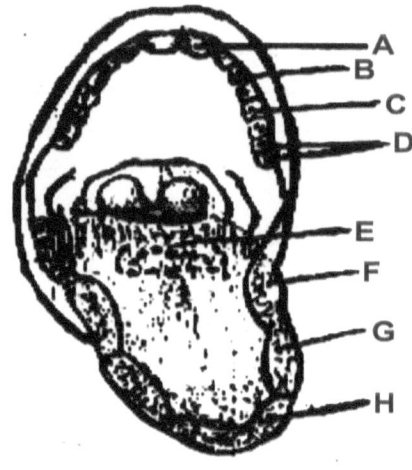

80A. Identify the teeth type in the diagram illustrated above.

A. _____

B. _____

C. _____

D. _____

80B. What are the shapes and functions of the teeth type in man?

_____

_____

80C. Define *dentition?*_____

80D. Mention the two main types of dentition and give examples of animals having them.

_____

_____

80E. Describe the adaptation of dentition of carnivore to its mode of feeding.

_____
_____
_____
_____

80F. Give the dental formulae of the following animals:

    i.   Man (permanent dentition):_____

    ii.   Dog:_____

    iii. Rabbit:_____

    iv. Horse:_____

    v.   Cow/Sheep:_____

    vi. Cat:_____

80G. Indicate the taste on the part of the tongue shown.

    E. _____

    F. _____

    G. _____

    H. _____

Study the diagram below and use it to answer questions A to G.

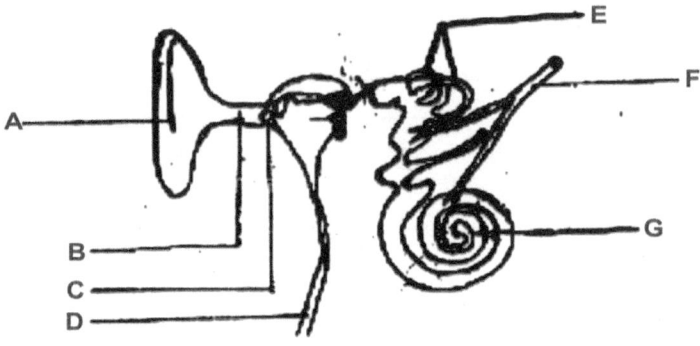

81A. Identify the structure illustrated above.

_____

81B. Give the two main functions of the structure illustrated above.

_____

81C. The structure is divided into three regions. Give the names of the regions.

_____

81D. Identify the structures labeled A to G.

A. _____
B. _____
C. _____
D. _____
E. _____
F. _____
G. _____

81E. Give the functions of the parts labeled A, B, C, D, and G.

A. _____

B. _____

C. _____

D. _____

G. _____

81F. Explain the mechanism of (i) hearing and (ii) balancing in the structure above.

_____

_____

81G. Mention the two main types of fluid in the structure above.

_____

Carefully study the different fruits illustrated below and answer questions A to F.

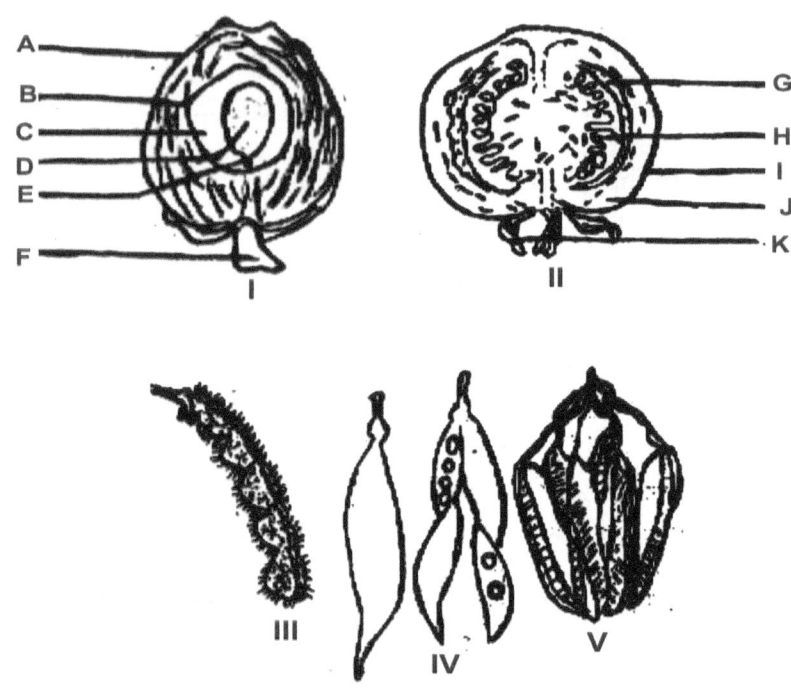

I   II   III   IV   V

82A. Identify specimens I, II, III, IV, and V.

I. _____

II. _____

III. _____

IV. _____

V. _____

82B. Give two reasons each for your answers.

_____

_____

_____

82C. Examine specimen III carefully.

    i.   State the agent of dispersal of specimen III:

    _____

    ii.   Give one structural feature of specimen III to support your answer.

    _____

82D. State two advantages of dispersal of fruits and seeds.

_____

_____

82E. Identify the parts labeled A to K.

A. _____

B. _____

C. _____

D. _____

E. _____

F. _____

G. _____

H. _____

I. _____

J. _____

K. _____

Study the diagram below and use it to answer questions A to D.

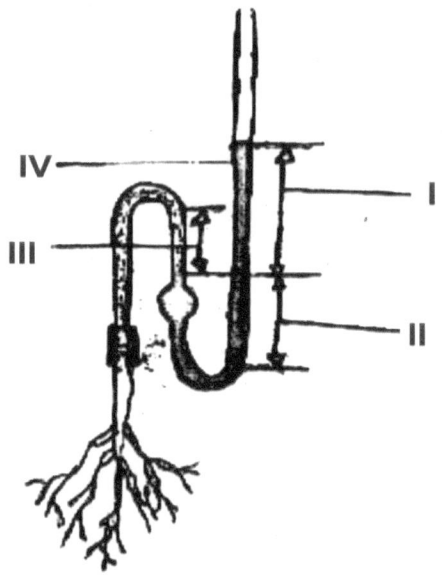

83A. Name the setup illustrated in the diagram.

_____

83B. What is the apparatus used for?

_____

83C. What is contained in the part of the apparatus labeled IV?

_____

83D. From which part(s) of the apparatus would the reading(s) be taken after the experiment?

_____

Identify the bones illustrated below and answer questions A to F.

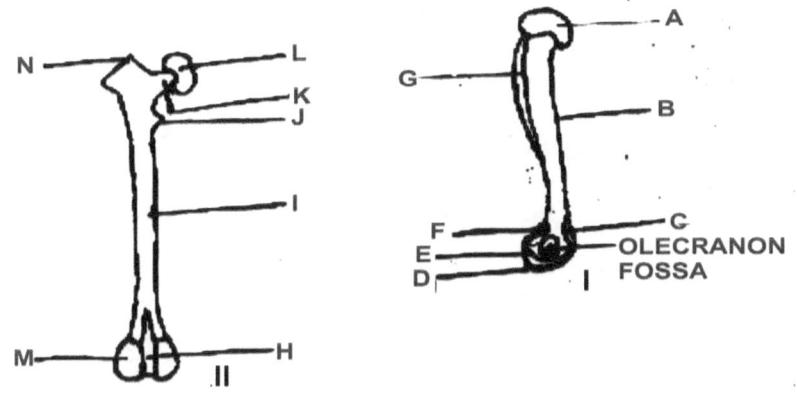

84A. Identify the bones illustrated above.

I. _____

II. _____

84B. Identify the parts labeled A to N.

A. _____   H. _____
B. _____   I. _____
C. _____   J. _____
D. _____   K. _____
E. _____   L. _____
F. _____   M. _____
G. _____   N. _____

84C. List two observable structural differences between diagrams I and II.

_____

_____

_____

84D. What are the functions of the two bones?

_____

_____

_____

84E. Name the bones that form joints with diagram II at the:

    i.   proximal end:_____

    ii. distal end:_____

Study the diagrams below and use them to answer questions A to F.

85A. Name the organisms which possess the mouth parts shown in diagram I and II.

    I.   _____

    II. _____

85B. Name the parts labeled A to I.

    A. _____   C. _____

    B. _____   D. _____

E. _____   H. _____
F. _____   I. _____
G. _____

85C. What types of feeding mechanism are the mouth parts in diagrams I and II adapted to?

I. _____
II. _____

85D. What types of feeding mechanisms are exhibited by mosquito, housefly, and tapeworm?

Mosquito: _____
Housefly: _____
Tapeworm: _____

85E. State one function each of the parts labeled G and H.

G. _____
H. _____

85F. What type of mouth part does the larva of the organism in diagram possess?

_____

The diagram below is a working model of the respiratory system of a mammal Study it carefully and use it to answer questions A and B.

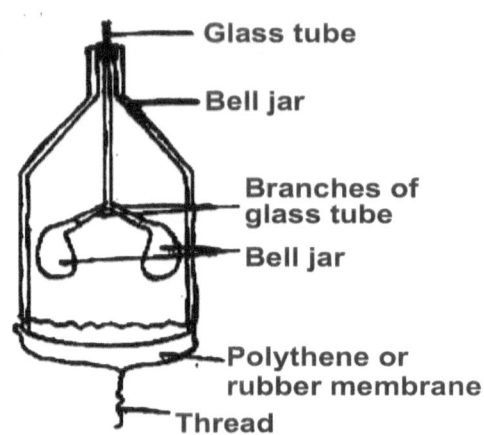

Glass tube

Bell jar

Branches of glass tube

Bell jar

Polythene or rubber membrane

Thread

86A. What do these labels represent in a live animal?

    i.   Glass tube:_____

    ii. Bell jar:_____

    iii. Branches of glass tube:_____

    iv. Balloons:_____

    v. Polythene or rubber membrane:_____

86B. Explain the behavior of the balloons...

    i.   When the rubber membrane is relaxed:

        _____

        _____

    ii. When the rubber membrane is pulled own:

        _____

        _____

iii. Use the working of the model to explain the process of inspiration of mammals

_____

_____

_____

87A. Examine the two types of germination illustrated in diagrams A and B and answer questions i to vii.

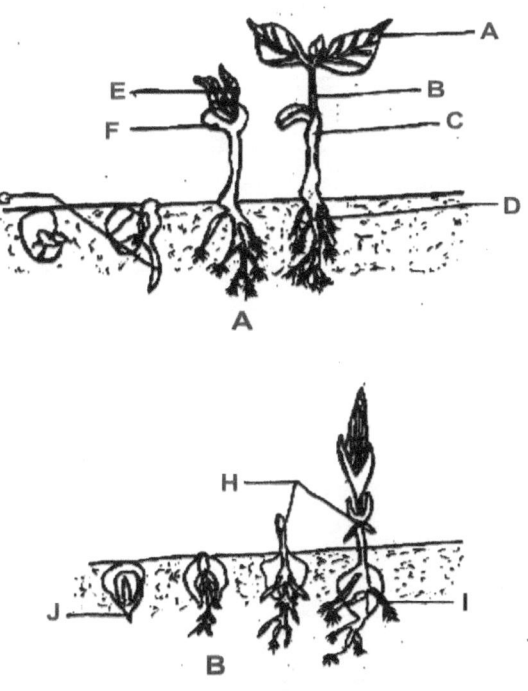

i. Mention the type of germination illustrated in diagrams A and B.

A. _____

B. _____

ii. Give three examples each of plants that have the same type of germination with A and B.

A. _____

B. _____

iii. Identify the parts labeled A to J.

A. _____   F. _____

B. _____   G. _____

C. _____   H. _____

D. _____   I. _____

E. _____   J. _____

iv. Mention five conditions that are necessary for germination to take place.

1. _____

2. _____

3. _____

4. _____

5. _____

v. Mention the type of root system in A and B.

A. _____

B. _____

vi. Which of the seedlings is a monocotyledon, and which is a dicotyledon?

_____

vii. In a tabular form, differentiate between a dicotyledon and a monocotyledon.

Dicotyledon	Monocotyledon

Study the diagrams below and use them to answer questions A to E.

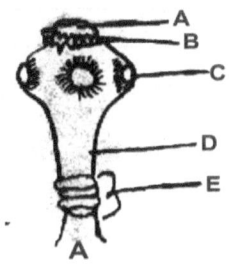
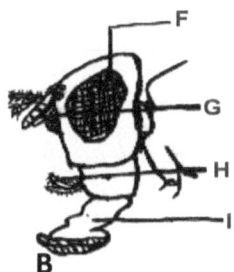

88A. Identify diagrams A and B.

A. _____

B. _____

88B. Name the parts labeled A to I.

A. _____   G. _____

B. _____   H. _____

C. _____   I. _____

D. _____

E. _____

F. _____

88C. What type of feeding mechanism are the mouth parts in diagrams A and B adapted to?

A. _____

B. _____

88D. What term is used to describe the sucking mechanism in diagram B?

_____

88E. What are the functions of the parts labeled A, B, and C in diagram A?

A. _____

B. _____

C. _____

Study the diagrams below. Use them to answer questions 89 (i) (v).

89A.

i. Give a general name for the structures illustrated in the diagram.

_____

ii. Identify the specimens illustrated in the diagram.

    I. _____
    II. _____
    III. _____
    IV. _____

iii. Identify the parts labeled A to F.

    A. _____   D. _____
    B. _____   E. _____
    C. _____   F. _____

iv. In which animal can the structures be found? Give the class and the phylum of the animal.

    _____
    _____

v. List three functions of the structures illustrated in the diagrams above.

    _____
    _____
    _____

90A. *Carefully study the diagrams illustrated below and answer question i to vi.*

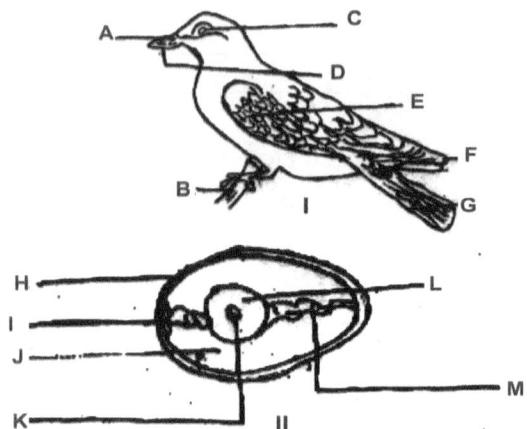

i. Identify I and II illustrated above.

I. _____
II. _____

ii. What is the relationship between I and II?

_____

iii. Identify the parts labeled A to M.

A. _____   J. _____
B. _____   K. _____
C. _____   L. _____
D. _____   M. _____
E. _____
F. _____
G. _____
H. _____
I. _____

iv. Give the functions of the parts labeled D, H, I, J, L, and M.

D. _____

H. _____

I. _____

J. _____

L. _____

M. _____

v.

    a.   Mention five kinds of feathers found in I.

    1. _____   4. _____

    2. _____   5. _____

    3. _____

    b.   Where are the feathers located?

_____

_____

_____

    c.   State the functions of the feathers.

_____

_____

_____

_____

Study the diagram below and use it to answer questions A to F.

91A. Identify the structure represented in the diagram above.

_____

91B. Mention the different layers/strata observable in the biotic community represented above.

_____

_____

91C. State five characteristic features of the community represented in the diagram.

_____

_____

_____

91D. Mention five climatic factors which affects the community illustrated above.

_____

91E. Mention five forest trees and five animals that can be found in the community.

_____

_____

91F. List five main biomes that characterize the Nigerian biotic community.

_____

_____

Carefully study the diagrams illustrated below and answer questions A to G.

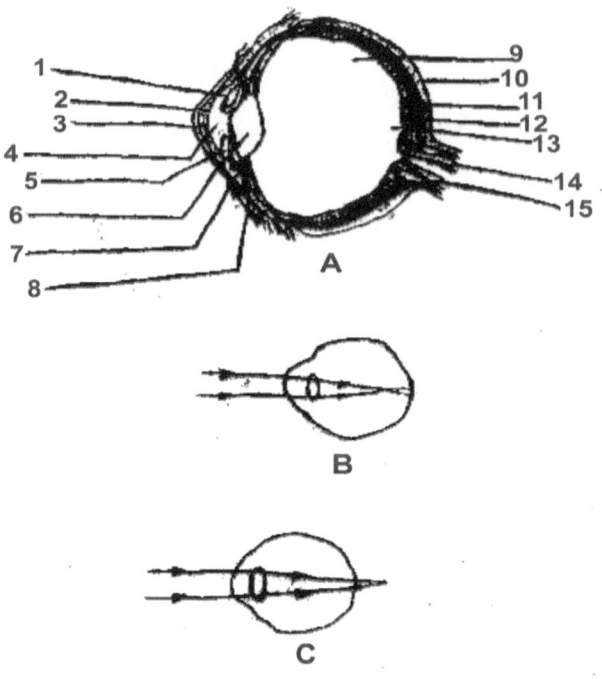

A

B

C

92A. Identify the eye defects illustrated in B and C.

B _____

C _____

92B.

i.   How can the defects be corrected?

_____

ii. State the effects of the correction in B and C.

B _____

C _____

92C. What is accommodation in relation to diagram A?

_____

92D. Identify the structure illustrated in A.

_____

92E. Identify the parts labeled 1 to 15.

1	_____	9	_____
2	_____	10	_____
3	_____	11	_____
4	_____	12	_____
5	_____	13	_____
6	_____	14	_____
7	_____	15	_____
8	_____		

92F. State the functions of the parts labeled 1, 5, 10, 12, 13, and 15.

1	_____
5	_____
10	_____
12	_____
13	_____
15	_____

92G. Give the similarities and the differences between human eye and the camera.

Similarities

_____

_____

Differences

Human Eye	Camera

Study the diagram below carefully and use it to answer questions A to H.

93A. Identify the structure illustrated above.

_____

93B. Classify the structure in to its kingdom, phylum, and class:

Kingdom:_____

Phylum:_____

Class:_____

94A.

    i.   Identify the diagram illustrated above.

    _____

    ii.   What is the setup used for?

    _____

    iii. Identify the parts labeled A to G.

        A. _____
        B. _____
        C. _____
        D. _____
        E. _____
        F. _____
        G. _____

iv. Give three precautions that must be taken in setting up the above apparatus.

_____

_____

94B. Two identical leafy shoots, C and D, from a shrub were each connected to the setup illustrated above. One was placed under normal laboratory conditions, and the other under a fan in the laboratory. The distance moved by the bubble in each setup was recorded at regular time intervals.

The results obtained are shown in the table below.

Time after start of experiment (minutes)	Distance moved by bubble under each environment condition (mm)		
	Shoot C: Under normal laboratory conditions		Shoot D: In the laboratory, under fan
0	0		0
5	4		7
10	6		12
15	9		20
20	12		30
25	14		35
30	18		40

i. Plot two graphs, one for each environmental condition, on the graph sheet provided with the distance moved by the bubble on the vertical y-axis and time on the horizontal x-axis.

GRAPH

Waec Standard Graph

ii.   At what time intervals were the readings taken?

_____

iii.   Under which condition did the bubble move faster?

_____

In testing for starch in a green leaf, a student took the following steps, as shown in the diagrams below. Carefully study the diagrams and questions A to E.

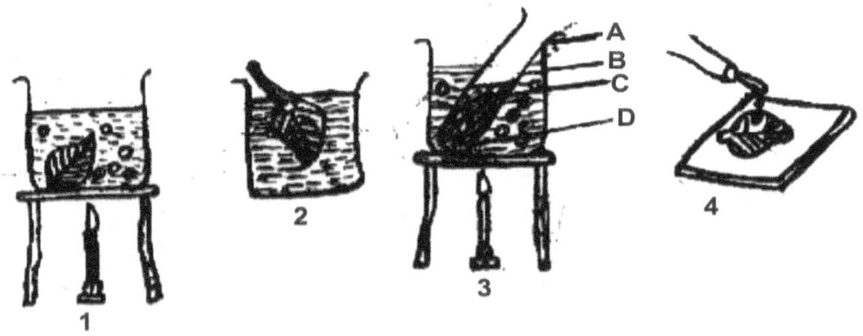

95A. Identify diagrams 1 to 4.

1. _____
2. _____
3. _____
4. _____

95B. Rearrange the steps in a sequential order.

_____
_____
_____
_____

95C. Give one reason each for steps 1, 2, 3, and 4 based on the diagrams illustrated above.

_____

_____

_____

_____

95D. Identify the parts labeled A to D.

A. _____

B. _____

C. _____

D. _____

95E. What color will indicate the presence of starch in the leaf?

_____

Study the diagrams illustrated below and answer questions I to VIII.

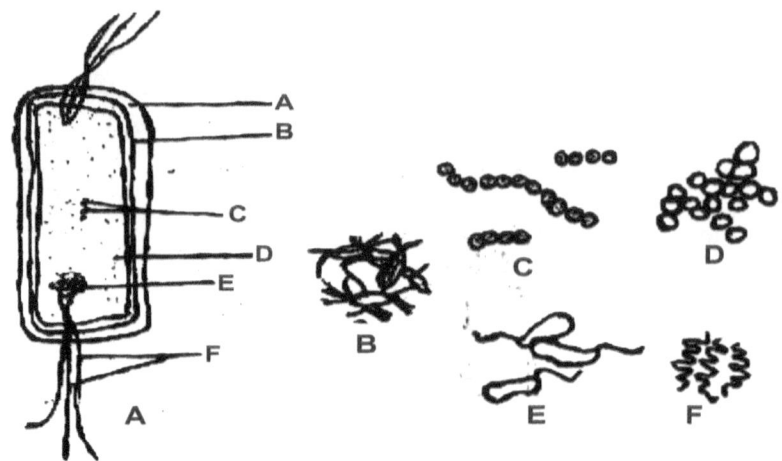

96A.

    i.   Identify Diagram A above.

    _____

    ii.   Identify the parts labeled A to F.

        A. _____   D. _____

        B. _____   E. _____

        C. _____   F. _____

    iii. Identify the different types of bacteria illustrated above.

        B. _____

        C. _____

        D. _____

        E. _____

        F. _____

iv.　List the diseases caused by B, C, D, E, and F.

B. _____

C. _____

D. _____

E. _____

F. _____

v.　Name five diseases each of plants and animals caused by Diagram A (other than the ones listed in above).

_____

_____

vi.　State five importance of Diagram A.

_____

_____

vii.　Give three ways in which Diagram A differs from the cells of other organisms.

_____

_____

viii. Where can the diagram illustrated in A above be found?

_____

Study the diagrams below carefully and answer questions A and B.

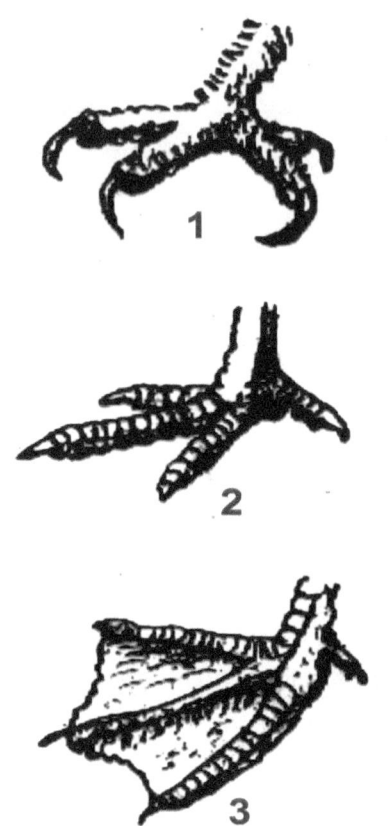

97A. Fill in the table below in respect of the diagrams.

Diagram	Type of Bird	Type of Feet	Special Uses
1	owl		
2		Strong feet, blunt nails	
3			for swimming quickly

97B. Use the diagrams below to fill in the table given.

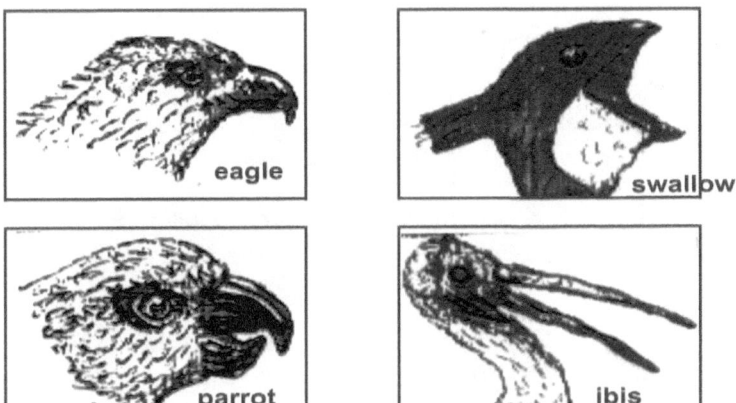

Type of Bird	Type of Beak	Special Uses
Eagle		
Parrot		
Swallow		
Ibis		

Study carefully the diagram below and answer questions A to E.

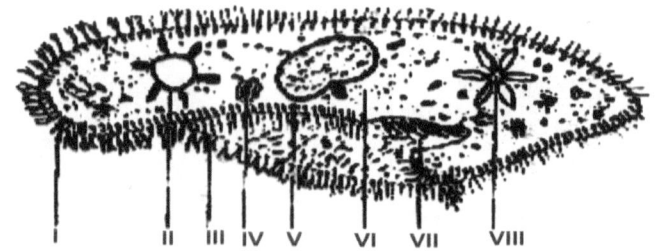

98A.

    i.   Identify the organism illustrated in the diagram above.

    _____

    ii.   Name the structures labeled I to VIII in the diagram.

I. _____	V. _____
II. _____	VI. _____
III. _____	VII. _____
IV. _____	VIII. _____

98B. State one function each of the structures labeled I, II, III, and IV.

I.   _____

II. _____

III. _____

IV. _____

98C. What will happen if the organism is placed in distilled water?

_____

98D.

    i.   To what level of organization does the organism belong?

    _____

    ii.   Give the reason it is placed in that level.

    _____

98E. Describe the mechanism of excretion in the diagram identified above.

_____

_____

_____

In a capture-release-recapture exercise to estimate the size of a population of dragon flies on a stretch of stream, 250 individuals were captured, marked, and released.

A second sample of 250 individuals was captured two days later, out of which 50 had the mark.

99A. Estimate the size of the population.

_____

_____

99B. Mention four precautions which should be taken in the capture-mark-release-recapture method.

_____

_____

99C. Briefly describe the use of a line transect in determining the population density of a land plant species.

_____

_____

_____

99D. Describe how the density of a plant species in an abandoned farmland could be determined.

_____

_____

_____

Study the diagrams illustrated below and use them to answer questions A and B

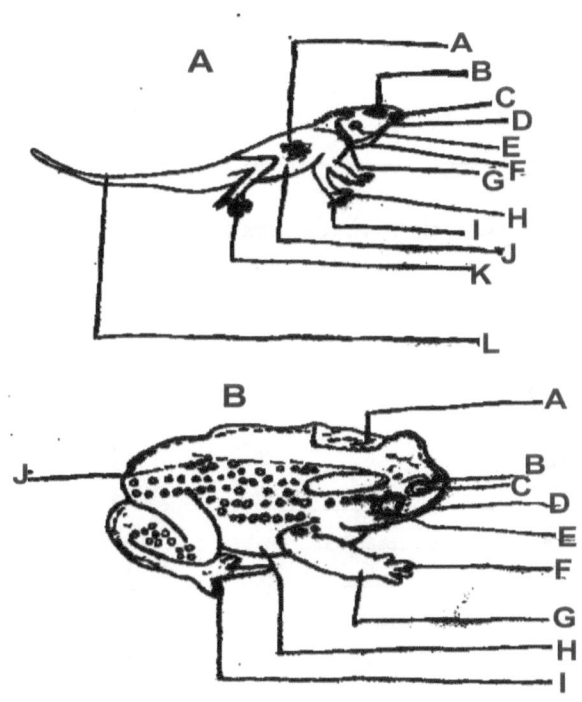

100A.

i. Identify the diagram illustrated above.

A. _____

B. _____

ii. To what groups of animals do they belong?

A. _____

B. _____

iii. State one external feature of Diagram A which is of evolutionary significance.

_____

iv. Comment on its significance.

_____

v. State two adaptive features of the limbs of Diagram A.

_____

vi. Identify the parts labeled A to L.

A. _____   G. _____
B. _____   H. _____
C. _____   I. _____
D. _____   J. _____
E. _____   K. _____
F. _____   L. _____

vii. Write out the male and the female features of the diagram illustrated in A.

Male Features	Female Features

ix. Briefly explain territorial behavior in the organism illustrated in A.

_____

_____

100B.

i. Identify the parts labeled A to K in the diagram illustrated
in B.

A. _____   G. _____
B. _____   H. _____
C. _____   I. _____
D. _____   J. _____
E. _____   K. _____
F. _____

ii. State, in tabular form, two differences between the forelimbs and hindlimbs of the diagram illustrated in B.

Forelimbs	Hindlimbs

iii. What is the importance of the differences between the
forelimbs and hindlimbs of the diagram illustrated in B?

_____

_____

iv. State the natural habitats of the organism illustrated in B.

_____

v. Differentiate between the organism illustrated in B and a frog.

_____

_____

ANSWERS

1A.

    A. Microscopic structure of *Paramecium/Paramecium caudatum*.

    B. Microscopic structure of *Amoeba/Amoeba proteus*.

    C. Microscopic structure of *Euglena/Euglena viridis*.

1B.   Names of the structures labeled 1–31.

1. Cilium
2. Pellicle
3. Contractile Vacuole
4. Ectoplasm
5. Food Vacuole
6. Trichocyst
7. Meganucleus/Macronucleus
8. Micronucleus
9. Mouth (Oral groove)
10. Gullet
11. Food vacuole forming
12. Endoplasm
13. Anal Pore
14. Collecting Canal
15. Pseudopodium
16. Nucleus
17. Plasma/Cell membrane
18. Ectoplasm
19. Endoplasm
20. Food vacuole
21. Food particle
22. Contractile vacuole
23. Flagellum
24. Eye spot
25. Paramylum-granules
26. Nucleus
27. Gullet

28. Reservoir
29. Contractile vacuole
30. Chloroplast
31. Thin living pellicle

1C.   Comparison of the three organisms:

	Paramecium	Amoeba	Euglena
Shape	Slipper-shaped	Shapeless	Spindle or pear-shaped
Food capture	Beating of cilia in oral groove, mouth, and gullet.	Food engulfed/ surrounded by pseudopodia.	Photosynthesis; absorption through cell membrane.
Site of food digestion	Food vacuole	Food vacuole	Cytoplasm
Absorption of food	From food vacuole by cyclosis and diffusion.	From food vacuole by cyclosis and diffusion.	From cytoplasm by cyclosis and diffusion.
Respiration	Inward diffusion of oxygen (O_2) and two outward diffusions of carbon(iv) oxide (CO_2).	Inward diffusion of oxygen (O_2) and two outward diffusions of carbon(iv) oxide (CO_2).	Inward diffusion of oxygen (O_2) and two outward diffusion of carbon(iv) oxide (CO_2).
Excretion	Via Cell Membrane	Via Cell Membrane	Via Cell Membrane
Reproduction	Binary fission conjugation	Binary fission conjugation	Binary fission conjugation
Sensitivity	Whole body responds to light, temperature, chemical, and contact.	Whole body responds to light, temperature, chemical, and contact.	Eyespot sensitive to light. Whole body responds to temperature and chemicals.
Movement	By heating of cilia.	By using pseudopodia and cyclosis (amoeboid movement).	By beating of flagellum and cyclosis (euglenoid movement).

1E. Function(s) of the parts labeled 6, 13, 20, 22, and 23:

Part labelled 6: Trichocyst. Contain filaments, which can be discharged, either to trap a prey and hold it, or use for offense and defense.

Part labelled 13: Anal Pore. A small opening through which undigested food material is egested out of the body.

Part labelled 20: Food vacuole. (This is formed when food is engulfed.) It provides site for the storage and digestion of food.

Part labelled 22: Contractile vacuole: It excretes waste liquid products such as dissolved ammonia. It regulates the amount of water and mineral salts that enter and leave the cell (osmoregulation).

Part labelled 23: Flagellum. For movement/locomotion.

1F. Phylum protozoa.

1G. Why the organism labelled C (euglena) is regarded as a plant:

i. The presence of chloroplasts used for photosynthesis.
ii. Definite shape (pear-shaped).
iii. Stores excess carbohydrates as starch (Paramylum granules are starch).

Only plants store starch. Animals store starch in form of glycogen.

1H. Where the organisms can be found:

A. Paramecium: found in freshwater containing decaying matter/muddy ponds/puddles/streams.

B. Amoeba: bottom of ponds/ditches/damp soil/mud/slow-moving streams.
C. Euglena: Non-flowing freshwater/ponds/ditches contaminated with excretion of farm animals.

2A.

i. Rat/rabbit/mouse/mammal/rodent.
ii. Naming of parts:
   I. External ear/pinna/earlobe
   II. Whisker
   III. Claw/digit
   IV. Forelimb
   V. Nipples (of mammary glands/teats)
   VI. Hind limb
   VII. Tail
   VIII. Hair/hair shaft/trunk/fur/skin
iii. Three observable features common with man:
   (External) ear/pinna—skin
   Hair on body—head
   Mammary glands—eye
   Forelimbs/arm—trunk
   Hind limb/leg—digits/toes

Two Differences between the Organism and Man

Organism	Man
Presence of whiskers	No whiskers
Presence of claws	No claws
Absence of nails	Presence of nails
Presence of tail	No tail
More than two mammary glands	Has two mammary glands

iv. Functions of the parts labeled II and V:
   II. Whisker for feeling the environment/to help avoid obstacles/danger during movement.

V. Mammary glands for the production of milk/to feed/ suckle the young.

3A.

i. Chlamydomonas
ii.
    I. Flagellum
    II. Anterior/apical papilla
    III. Contractile vacuole
    IV. Eyespot/Stigma
    V. Cytoplasm
    VI. Nucleus
    VII. Cell wall/Cellulose cell wall
    VIII. Chloroplast
    IX. Pyrenoid

3B.

i. A Plant/Plantlike
ii. Reasons: presence of Chloroplast/Chlorophyll.

Possession of definite shape and possession of cellulose cell wall/cell wall.

Possession of pyrenoid for the storage of starch.

3C. Functions of labeled parts:

    V. Contractile vacuole: elimination of excess water/ excretion/osmoregulation.
    VI. Eyespot/stigma: sensitive to light/irritability.
    VII. Nucleus: for reproduction/controls all cell activities.

3D. The organism: Chlamydomonas is regarded as primitive because of the flagellum used for movement/locomotion

because the organism is unicellular/unicell at the cellular level
of organization.

4A.

A. Liver
B. Gizzard
C. Small intestine
D. Large intestine/colon
E. Salivary gland
F. Stomach/ventriculus/midgut/mid intestine/mesenteron
G. Malpighian tubule
H. Ileum/hindgut/hind intestine
I. Colon/hindgut/hind intestine
J. Gastric/midgut/caecum
K. Crop
L. Pancreas
M. Proventriculus/stomach
N. Beak

4B. Functions

B. For crushing/grinding (food)
J. For production/secretion of digestive enzyme
K. For storage (of ingested food)
M. Secretes (digestive) enzymes

Adaptive features:
B. Has thick (muscular) wall/rough wall/contains stones/
   pebbles
J. Creates large surface area (for secretion and storage of
   digestive enzyme)
K. Large/saclike
M. Glandular/contains glands

4C. Similarities: They both have:

Crop
Muscular gizzard
Small and large intestines
Narrow Esophagus
Anus
Hollow tubes that open at both ends
Caecum

Differences

Diagram I (Bird)	Diagram II (Cockroach)
Salivary gland: absent	Salivary gland: present
Liver: present	Liver: absent
Pancreas: present	Pancreas: absent
Small intestine: very long	Small intestine: short
Malpighian tubule: absent	Malpighian tubule: present
Gastric caeca: absent	Gastric caeca: present
Hindgut terminates in cloaca	Hindgut terminates in anus

4D.

i. Type of food fed on:
    Seed(s)/grain(s), fruits, insects, worms, leaves, foliage, or vegetables.
ii. Reasons:
    Short beak (for picking and cracking seeds)
    Long-coiled alimentary canal (indicating herbivorous diet/for digestion of plant materials)
    Gizzard (for crushing grains, etc.)

5A. The organism is tadpole/larva of toad/frog.

5B.

    I. External gill(s)
    II. Mouth
    III. Operculum/gill cover
    IV. Tail muscle/myotome
    V. Tail fin/fin

5C. For respiration/gaseous exchange/breathing.

5D. Freshwater/aquatic habitat/pond/swamp/lake/slow-running water.

5E.

    i. Scrapes/nibbles on plants in water
    ii. By swimming/using the tail fin/brought about by the contraction of myotome /tail muscle

6A.

    A. Mouth/horny jaw
    B. (V-shaped) cement gland
    C. External gills/gills
    D. Alimentary canal/intestine
    E. Tail/Tail fin

6B.

    A. Honey jaw: for chewing, nibbling, scraping vegetable/plant material.
    C. External gills/gills: for breathing, respiration, gaseous exchange.

6C. Internal gill stage.

6D.

    F. Egg
    G. Jelly/jelly-like substance/jelly-like string.

6E. Functions of G

Protects the egg (from bacterial infection/animal attack).

Keeps eggs afloat/attaches eggs to weeds.

Separate eggs from each other/allows room for entry of oxygen/prevents overcrowding of the eggs.

Prevents eggs from drying up.

6F. The structure labeled F/egg develops/hatches into tadpole/diagram I

6G. The egg contains yolk/food storage for nourishing the embryo.

Has protective covering against mechanical injury/drying up.

7A.

    A. Chiasmata
    B. Exchange of genes/genetic/hereditary/chromosomal/segments/parts/materials which leads to variation/mutation.
    C. Late prophase (I), Metaphase I/Diplotene

8A.

    A. Asexual reproduction/vegetative reproduction

8B. A Budding

   B. Grafting

8C. A Yeast/hydra

   B. Mango/Orange
   C. Bryophyllum/Kalanchoe/Begonia/Amoeba/Euglena/ Chlamydomonas/Paramecium

8D.

   i. Advantages:
       1. More offspring are produced within a short time.
       6. New plant obtains nourishment from parent plants.
       7. New plants produced mature faster.
       8. Asexual reproduction does not depend on external agents of pollination; only one organism is involved.
       9. New plants resemble their parents/no variations in the offspring/perpetuation of species.

   ii. Disadvantages
       1. Reduces the strength and vigor of the succeeding generation.
       2. Undesired qualities/diseases are transmitted to offspring.
       3. No variations in offspring.
       7. Offspring may not withstand environmental changes
       8. There can be overcrowding/competition among offspring/only few organisms survive.

9A.

   A. Esophagus
   B. Stomach
   C. Pancreas

D. Pancreatic duct
E. Duodenum
F. Bile duct
G. Gall bladder
H. Liver

9B. Liver/liver lobes/liver cells.

9C. Pancreas

9D. Insulin

9E. There will be less or no bile production; therefore, rate of fat digestion will be reduced/less, or no fat emulsification.

10A. At point A, the CO_2 concentration is slightly higher than O_2 concentration, hence the pollution is slight.

At point B, the CO_2 concentration is much higher than that of O_2 (27–27 ppm); hence, there is much/heavier pollution.

10B. At point A, the pH is higher (than at B), which means lower acidity (i.e., 19 ppm CO_2); it therefore accounts for higher population density of two organisms.

At point B, pH is lower than at A, which means higher acidity (i.e., lower population density).

10C. Oxygen concentration is higher at point A (in fact double) than at point B.

Lower pH/higher acidity at B affects survival of organisms. Heavier pollution at point B than at A.

10D. Organisms that have high tolerance for low oxygen concentration [e.g., algae, fungi/anaerobic organisms (bacteria)].

11A.

    A. Lid/cover
    B. Pitcher
    C. Hook
    D. (Ventral) sucker
    E. Proglottid/proglottis

11B.

    I.   Carnivorous mode of feeding/insectivorous
    II. Parasitic mode of feeding

11C.

Tapeworm	*Rhizopus*
No digestion required. Feeds on digested food of a saprophytic living host/parasitic.	(Extracellular) digestion required. Saprophytic.

11E. Insects, spider, ant, etc.

11F.

Presence of hook(s)
Sucker(s)
Flat body
Detachable proglottid/segment

11G.

I. The pitcher has a lid/cover. The lid undergoes nastic movement closest in response to the touch, pitcher secretes enzymes, insect is digested, and the pitcher absorbs the nitrogenous compound/digested food.

II. It attaches itself to the host's intestine with the aid of hooks and suckers and absorbs the digested food of the host through the entire body surface.

11H. Intestine of man/mammal/dog/sheep/fish/rat/bird/cat/pig.

11I.

Anemia
Weakness
Malnutrition/always hungry/persistent hunger/under nourished
Loss in weight
Shortage of blood
Emaciation

12A.

I. Chlamydomonas
II. Volvox

12B. I & II Cellular level of organization

12C. I is a unicellular organism.

II is a colonial alga.
  I. It is a unicellular organism. It is made up of mass of protoplasm enclosed in a membrane with organelles that can carry out all life processes.

II. It is a colonial alga. Each cell of the colony exists independently.

13A.

i. Longitudinal/vertical section
ii. Horizontal/transverse section/cross section

13B.

A. Epicarp
B. Mesocarp
C. Endocarp
D. Seed/cotyledon/seed leaf
E. Remains of style
F. Pericarp/fruit wall
124

13C.

i. Differences between Fig. I and Fig. II

Fig. I	Fig. II
Has one seed (drupe)	Has more than one seed (berry)
Endocarp is thinner; mesocarp is wider	Endocarp is wider; mesocarp is thinner
Strong/woody/hard endocarp	Fleshy/pulpy/juicy endocarp
Has a large seed	Has small seeds
Seed with soft, thin testa	Seed with hard testa
Formed from monocarpellary ovary; not divided	Fruit divided into sectors/chambers/loculi formed from syncarpous ovary

ii. Similarities Between Fig. I and Fig. II
They have seed(s).

Both have fruit walls; pericarp is three-layered, divided into epicarp, mesocarp, and endocarp.

Both are fleshy/succulent.

13D. Differences between Fig. I and a typical seed

Fig. I	Fig. II
Has two scars (i.e., those of fruit's stalk and style)	Has one scar (that of funicle)
Covering of Fig. I is fruit wall/pericarp; outermost covering is the epicarp	Covering of seed is the seed coat or testa; outermost covering is testa or seed coat.
Has fruit wall made up of three layers	Has seed coat made up of two layers
Has a fruit wall	Has no fruit wall
Has no seed coat	Has seed coat

13F. Color of orange/ripe fruit attracts the man (animal).

Fleshy endocarp eaten consumed by man.
Seeds thrown away.

13G.

    G. Oil glands
    H. Epicarp
    I.   Mesocarp
    J.   Placenta
    K.   Seed
    L.   Juicy endocarp
    M. Succulent hair attached to endocarp

14A.

Savannah/forest/pond

14B.

A. Grass
B. Grasshopper/locust/insect
C. Tree/shrub/stem
D. Toad/frog
E. Waterweed/water plant
F. Water
G. Anthill/termitarium
H. Termite
I. Fish

14C. Grasshopper/locust, toad/frog, termite, fish

14D. The green plants absorb solar energy from the sun:

For photosynthesis; to build up/synthesize plant proteins which contains chemical energy; the herbivores (e.g., insects) feed on green plants; they are, in turn, eaten by toad; the toad is eaten by big fish; and for synthesis of animal protein which contains chemical energy.

14E.

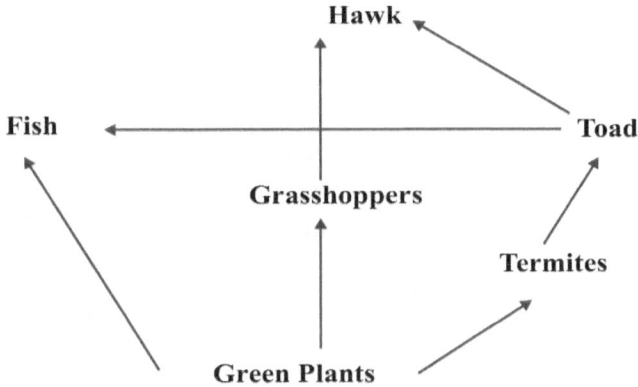

14F. Bacteria or any named bacteria:

Fungi

14G. Grass

15A.

To demonstrate the water culture experiment/to demonstrate the (mineral) elements essential for healthy plant growth, to show that plants need some (mineral) elements for healthy growth, and to demonstrate the effects of the deficiency or absence of certain (mineral) elements on plant growth.

15B.

A. Vent/aeration tube/bent-glass tubing/tube
B. Plant/seedling/stem (of plant seedling)
C. Cork/rubber bung/stopper
D. Dry cotton wool
E. Water culture solution or Knop's/Sach's solution
F. Wide-mouth bottle/container/jar

15C. Role of A, D, and E:

A. For aeration (of the culture solution) to allow in air/oxygen.
D. To hold seedlings in place, to keep the seedling dry and prevent growth of microorganisms on the stem, or to protect seedlings from damage.
E. Contains/provides the nutrients needed for plant growth/contains the nutrients under investigation.

15D. Precautions

Apparatus must be very clean/sterilized to prevent infection of seedling/contamination of nutrient solution (by trace elements).

Vessel should be painted black or covered with black cloth/paper/aluminum/silver foil to prevent algal growth to cut off light from the culture solution (which will damage the roots).

Constant aeration (of the culture) to provide oxygen/air for the root.

Stopper or cotton wool must be kept dry so that the stem of the seedling enclosed are not wet. Wetness increases risk of microbial infection of the seedling.

Frequent topping-up/ replacement of the culture solution (every fortnight) to ensure adequate supply of required nutrient.

15E. Functions of elements:

Nitrogen: protein synthesis/synthesis of chlorophyll (Chloroplast), constituent of all proteins/enzymes/protoplasm, Nucleic Acid Synthesis.

Phosphorus: formation of nucleic acid, present in enzymes/ co-enzymes, important in nuclear division, and acts as buffer in cell sap; stem and root formation.

Calcium: Cell wall formation, healthy growth, neutralizes certain organic acids, activates certain enzymes, and gives rigidity to plant

15F. Macroelements: these are elements required/needed in large quantities for healthy growth of plants.

Microelements: these are elements required/needed in very small/minute quantities for healthy growth of plants.

16A.

A. Pericarp/fruit wall/pericarp and testa/seed coat fused
B. Endosperm
C. Scutellum/cotyledon
D. Coleoptile/plumule sheath
E. Plumule
F. Radicle
G. Coleorhiza
H. Remains of style

16B. Monocotyledonous plant/monocotyledon

16C. Endosperm

16D. Water is absorbed (into the endosperm).

This causes the grain to swell up.

Starch and protein are digested by enzymes (activated by water).

The foods (in soluble form) are absorbed into the scutellum and transported to the embryo/radicle and plumule.

The radicle elongates/comes out (along with the coleorhiza); the radicle bursts through the coleorhiza/grows in to the soil downward.

The plumule grows upward, but the rest of the grain remains in the soil (i.e., hypogeal germination).

17A.

    I.   Incisor
    II. Canine
    III. Premolar
    IV. Carnassial (teeth)
    V.   Molar

17B. Carnivorous (mode of feeding). Feeds on flesh.

17C.

    I.   Incisor: small pointed or chisel-shaped—used for prehension/seizing/holding prey/food.
    II. Canine: long, curved, and pointed—used for holding or seizing pray/tearing flesh.
    III. Premolar: have blunt cusps (surface)/cutting edges—used for cutting food.
    IV. Carnassial teeth: enlarged with sharp edges for crushing/grinding bones/tearing flesh from bones.

17D. Dog/cat/lion/leopard/hyena/tiger/fox/civets or any other correctly named carnivore.

18A.

    A. Kidney
    B. Ovary
    C. Intestine/large intestine/stump of intestine
    D. Uterus/womb
    E. Vagina
    F. Anus
    G. Urethra
    H. (Urinary) bladder
    I. Fallopian tube/oviduct/fallopian funnel
    J. Ureter

18B. Female

18C. Kidney, Urethra, (Urinary) bladder, ureter

18D. Ovary

18E. Intestines, Anus

19A.

    A. Adrenal gland/body
    B. Kidney
    C. Ureter
    D. Ovary
    E. Fallopian tube/oviduct
    F. Wombs/uteri or uterus
    G. Vagina
    H. Rectum
    I. Vulva
    J. Anus

K. Urethral aperture/opening/urethra
L. (Urinary) bladder
    iv. Female
    v. Presence of womb/uterus
       Presence of vagina/vulva
       Presence of ovary
       Presence of fallopian tube/oviduct

19B. There are two uteri/wombs in the diagram, but only one uterus/womb in the human being.

19C. Functions

E. Oviduct: Transportation/passage for ovum/egg/zygote to the uterus/site for fertilization.
F. Womb: Womb is where the fetus/embryo develops/site of implantation of the zygote.
For nourishment of the fetus.
G. Vagina: Mating/copulation/receives sperms from male/ serves as a birth canal during the expulsion of the fetus from the uterus.
L. Bladder: Temporary storage of urine.

20A.

i. Musa is AS and Obiageli is AS
ii. Genotype of Musa's father: AA
iii. Genotype of Obiageli's father: SS

20B. Musa and his wife (Obiageli) are carriers of the gene for sickle cell disease/Musa and Obiageli have the AS genotype.

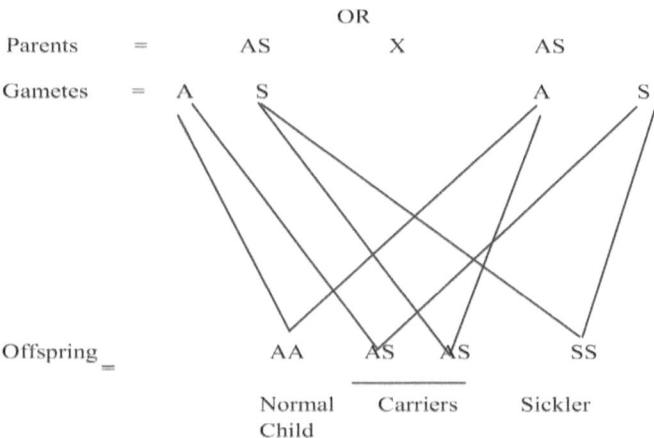

20C. Ada and Abiola could have genotypes AA or SS.

20D. Advise Ada to choose a man who is neither a sickler (i.e., SS) nor carrier of the sickle cell gene (i.e., AS).

Advise Ada to marry a man that is homozygous for normal hemoglobin (i.e., AA).

21A. Savanna/guinea savanna/grassland

21B. Tall grass/plenty of grass/scattered trees

Trees with broad leaves
Thick bark
Fire resistant trees abundant
Many herbivores
Carnivorous animals present

21C. Humidity, temperature, wind, light, soil pH, sunlight, atmospheric pressure

21D. Humidity: hygrometer

Temperature: thermometer
Wind direction: wind vane
Light: photometer
Soil pH: pH meter
Pressure: barometer
Wind speed: anemometer

21E.

  i.   Less tall tress/few trees that are hardy, plenty of grass.
  iv. After some rain the grass and shrubs, which had either died or lost their leaves, are now in full leaf, the grass grows up in tufts and almost covers the bare ground.

21F.

  i.   Plants: shea butter, baobab, elephant grass, few xerophytes, locust bean trees, acacia, etc.
  v.   Animals: herbivores   (antelopes/sheep/rodent/goat/grasscutter/elephant/monkey).
       Carnivores (i.e., lion, rodents, termites, weaverbird, quelea bird, lizards, snakes, etc.).

21G.

  i.   Plants: thick bark, small leaves
  ii. Animal: muscular legs.
       Sharp claws and teeth
       Chewing the curd

21H. Adaptation:

    i.   Plant: thick bark to resist fire/conserve water.
        Small leaves to conserve water/reduce transpiration

    ii. Animal: muscular legs to run fast/to run away from predators/to catch prey.
        Sharp claws and teeth for grasping or catching prey.
        Chewing the curd to graze quickly/feed quickly and regurgitate at a safe corner/place.

22A. Mitosis/cell division/mitotic cell division

22B. Plants: root tip/terminal bud/developing organ/cambium/meristems

Animals: in the skin (germinative layer) binary fission in amoeba, bone marrow, spleen, liver—all correctly named organs/structure (e.g., stomach, esophagus, ileum).

22C.

    A. Centriole/spindle
    B. Chromosomes/chromatids
    C. Nuclear membrane/nucleus
    D. Cytoplasm

22D. Important in growth/development.

Important for repair of damaged cells/organs.

For maintenance/retention of the chromosome number (from generation-to-generation)/for maintenance of the character or form of the species from generation to generation. Or ensuring that the diploid condition of the cells is retained.

Important for asexual reproduction in unicellular organisms.

22E. Stage I/stage II/prophase

22F. 46 Chromosomes

23A.

    A. There is interaction between biotic and abiotic factors/living and nonliving things; the sun and the rain (abiotic factors) interacting with the (biotic factors) living things—insects, grass, bird, cattle, etc.

    B. Cattle egret feeds on parasites on the cattle.
Cattle, on its part, gets its parasites removed by the egret.
During grazing, the cattle stirs up insects for egret to feed on.

    C. Symbiosis/symbiotic association/mutualism

    D. Green plant/grass→Grasshopper→Hawk
Grass/Plant→Earthworm→Egret/Hawk
Grass/Plant→Rat→Hawk
Grass/Plant→Earthworm→Small bird→Hawk
Humus/decaying matter→Mushroom→Hawk
Grass/Plant→Small birds→Hawk

    E. There will be overgrazing.
Leading to shortage of food/competition for food for the animals (rats/grasshopper etc.)
Resulting in reduction in their population.
Erosion may occur, leading to poor soil/destruction of soil/low organic matter in the soil.
Hence, poor plant growth.

24A. Hydrophyte/water lettuce/Pistia

24B. Light body

Poorly developed strengthening tissue.

Presence of tissues with air spaces/spongy tissue/presence of aerenchyma.

Presence of poorly developed conducting tissue.

Presence of thin cuticle/no cuticle.

Presence of stomata only on the upper surface of the leaf—presence of stolon.

Presence of fibrous roots for water absorption.

25A. To demonstrate Osmosis in a living tissue/experiment to demonstrate that water moves from region of higher concentration to region of lower concentration through a semipermeable membrane.

25B. The liquid labeled I is water.

25C. Peeled yam, potato, pawpaw, pig's bladder, cocoyam.

25D.

    i.   The level of the solution in A will rise, while the level of B will remain the same.

    ii.   In setup A, the trough has a higher concentration of water molecules than the sugar solution.

        So there is a movement of water molecules from the trough through the living tissue into the sugar solution.

        There is a higher osmotic pressure in the sugar solution than in the trough, so water molecules move from the trough through the cells of the living tissue into the sugar solution; *or* here is a higher osmotic pressure in the sugar solution than in the trough, so water molecules move from the trough through the cells of the living tissue into the sugar solution.

In setup B, the concentrations of water molecules are the same in the trough and in the yam cup: no movement of water molecules.

25E. Setup B serves as a control.

26A.

    i.   Transverse/horizontal/cross section of a dicotyledon stem/ dicotyledonous stem.

    ii. Reasons for answer:
        Presence of vascular bundles arranged in a ring.

        Presence of interfascicular cambium/complete ring of cambium.

        Each vascular bundle distinctly differentiated into phloem, cambium, and xylem.

        Presence of pith/pith forms widest tissue.

        Differentiation of cortex into collenchyma and parenchyma tissues.

        Presence of epidermis.

26B.

    I.   Epidermis
    II.   Collenchyma
    III.   Parenchyma
    IV.   (Interfascicular) cambium
    V.   Phloem
    VI.   Xylem
    VII. Pith
    VIII. Cortex

26C. Functions of I (epidermis):

Forms the protective tissue of the stem (i.e., for protection)
Waterproof/checks excessive loss of water
Allows exchange of gases (through lenticels)
Prevents entry of dust and microorganisms (fungi/bacteria)

Functions of V (Phloem):
Translocation/conduction/transportation of manufactured food
Storage of food
Production of latex/sap
Some portion may perform supporting tissue
Function of VI (Xylem):

Transportation of water and mineral salts up the stem
Forms a supporting tissue
Function of VII (Pith):

May perform strengthening functions/supportive function.
Storage of foods.
Air-space system allows air to circulate within the living region of stem.

26D. Significance of parts II and III

    II. Collenchyma support/strengthening
        Can perform photosynthesis
        Provides resilience and flexibility to plants
    IV. Parenchyma
        For support/strengthening
        Can perform photosynthesis
        Can store food and water
        Allow air to diffuse among the cells

26E.

    i.   Formation of the part labeled IV (cambium) leads to:
          The process of secondary thickening
          Formation of secondary xylem/wood
          Formation of secondary phloem
          Formation of bark/cork/lenticles

    ii. Importance of secondary thickening:
          Strengthening of stem/supportive function
          Increased rate of conduction of water/mineral salts
          Increased rate of conduction of food
          Increase in girth/thickness
          Increase in protection against biodeterioration/microbial decay
          Formation of bark/cork to protect the inner tissue of stem
          Formation of lenticels which allow exchange of gases

26F.

    i.   Reagent for identifying the tissue is iodine solution
    ii. Color to observe is blue-black/blue-black, and the substances indicated by the color is starch.

27A.

    I.   Cortex
    II.   Branch of renal artery or vein/renal artery or vein
    III. Pelvis
    IV. Pyramid
    V.   Ureter
    VI. Medulla

27B. Functions of II and V

    II. Carries (oxygenated) blood/nutrient (from aorta) to the kidney; carries (deoxygenated) blood away from the kidney.

    V. Carries urine to the bladder (urinary bladder/carries urine from the kidney).

27C. Liver

28A.

    I. Three possible modes of nutrition:
Autotrophic/holophytic/photosynthetic
Saprophytic/saprobic/saprophytism
Holozoic/heterotrophic

    II. Reasons
Autotrophic/holophytic/photosynthetic because of chlorophyll or green pigment present in green plants.
Saprophytic/saprobic because of the mushroom (fungus) present.

28B. Simple nitrogen cycle

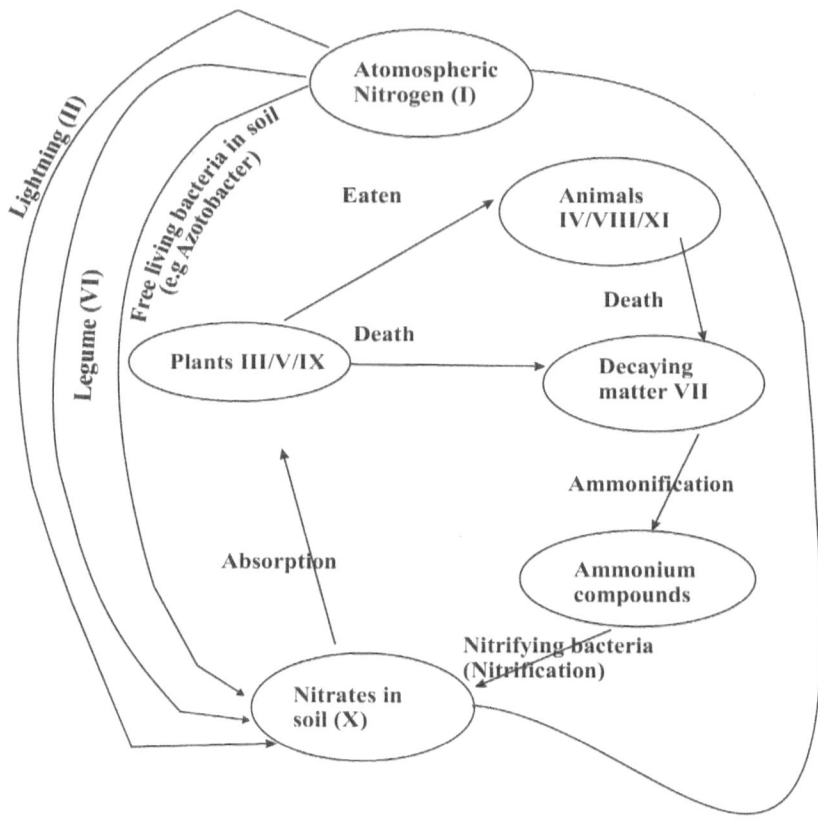

28C. Explanation of processes involving

I. I, II, and X
    During thunderstorm, lightning (II) releases energy which combines atmospheric nitrogen (I) with oxygen to form nitrogen dioxide/nitrous oxide/oxides of nitrogen.

    These gases dissolve in raindrops to form nitric acid (HNO_3) in the soil (X).

Nitrogen dioxide/Nitric acid combines with metallic parts of salts to form nitrates.

II. IV, VIII, and X.
Snake (IV) kills rats (VIII)/snake feeds on rat

Dead rat/snakes acted upon by the putrefying bacteria

To cause their decay—decaying matter—(VII) to form ammonium compounds which are acted upon by nitro-bacter/nitrosomonas/nitrifying bacteria in the soil (X) to form nitrates.

28D.

i.
  1. Mushroom/fungus/fungi
  2. Earthworm
ii. Drawing of food chain
Green plant (III)→Rat (VIII)→Snake (IV)
Mushroom→Rat (VI)→ Snake (VIII)→(IV)
Weed (IX)→Rat→(VIII)→ Snake (IV)

28E.

I. Ecosystem
II. Biotic factor = animals/snake/rat/earthworm/plants/mushroom
Abiotic factors = atmosphere, air, lightning, nitrogen, soil, rainfall

29A. To demonstrate suction pressure/transpiration pull

29B. To prevent air bubbles from blocking/entering the xylem vessels

29C. Secchi disc

30A.

A. Salivary gland	N. Appendix
B. Mouth	O. Colon/large intestine
C. Esophagus	P. Rectum
D. Cardiac sphincter	Q. Anus
E. Gall bladder	R. Beak
F. Liver	S. Esophagus
G. Stomach	T. Crop
H. Spleen	U. Stomach
I. Pancreas	V. Duodenum
J. Pancreatic duct	W. gizzard
K. Duodenum	X. Caecum
L. Small intestine/ileum	Y. Small intestine
M. Caecum	Z. Cloaca (anus)

30B. Digestion of starch:

The food is chewed /masticated.

Saliva is secreted by the salivary glands.

Saliva contains the enzymes ptyalin/salivary amylase.

The saliva provides an alkaline medium for ptyalin/salivary amylase to work.

The ptyalin acts on starch converting it to maltose.

Amylase in the pancreatic juice converts the remaining starch to maltose.

Maltase converts the maltose to glucose in the small intestine.

Digestion of protein:

The food is chewed, mixed with saliva, and swallowed.

The stomach secrets dilute HCl (dilute hydrochloric acid) and gastric juice.

Gastric juice contains pepsin and rennin.

The contents of the stomach are then churned to form chyme.

Dilute hydrochloric acid (dilute HCl) provides acidic medium for the pepsin and rennin to work.

Rennin coagulates milk proteins: pepsin digests protein to polypeptides.

The cells of the pancreas also make enzymes (i.e., trypsin) which break down proteins to peptides and peptides to soluble amino acids.

30C. Significance of the long coiled structure labeled L (i.e., small intestine).

To provide a large surface area for digestion and absorption of food substances.

30D. Similarities between the digestive system of man and that of a bird:

Both possess esophagus, small intestine, large intestine, pancreas, liver, spleen, tongue, rectum, anus.
Differences between the digestive system of man and that of a bird

Man	Bird
1. Gizzard absent	Gizzard present
2. Mouth with teeth	Mouth modified into beak
3. Crop absent	Crop present
4. Appendix present	Appendix absent
5. Epiglottis present	Epiglottis absent
6. Stomach present	Stomach absent
7. Proventriculus absent	Proventriculus present

30E.

I.   Functions of the parts labeled N and R:
N.   Appendix: It has no known function in man, but it is a notorious seat of the disease called appendicitis.
R.   Beak: For picking food materials
      For cracking seeds
      For killing prey and tearing off strips of flesh
      Used in climbing (in some birds)
      For probing earth and mud
      For gripping slippery animal
      For spearing fish
II. An inflamed (N) appendix is surgically removed

31A.

i.   Reasons for population decrease in specimen A after twenty-five days:

Species B feeding on species A/predation.

Depletion/exhaustion/scarcity of food.

Death of species A (due to aging).

Decline in reproduction/breeding (due to aging).

ii. Why population of species B decreased after forty days

Competition among species B, resulting in survival of the fittest

Scarcity/exhaustion of food/prey/species A

Death of species B (due to aging)

Decline in reproduction/breeding in species B (due to old age)

GRAPH

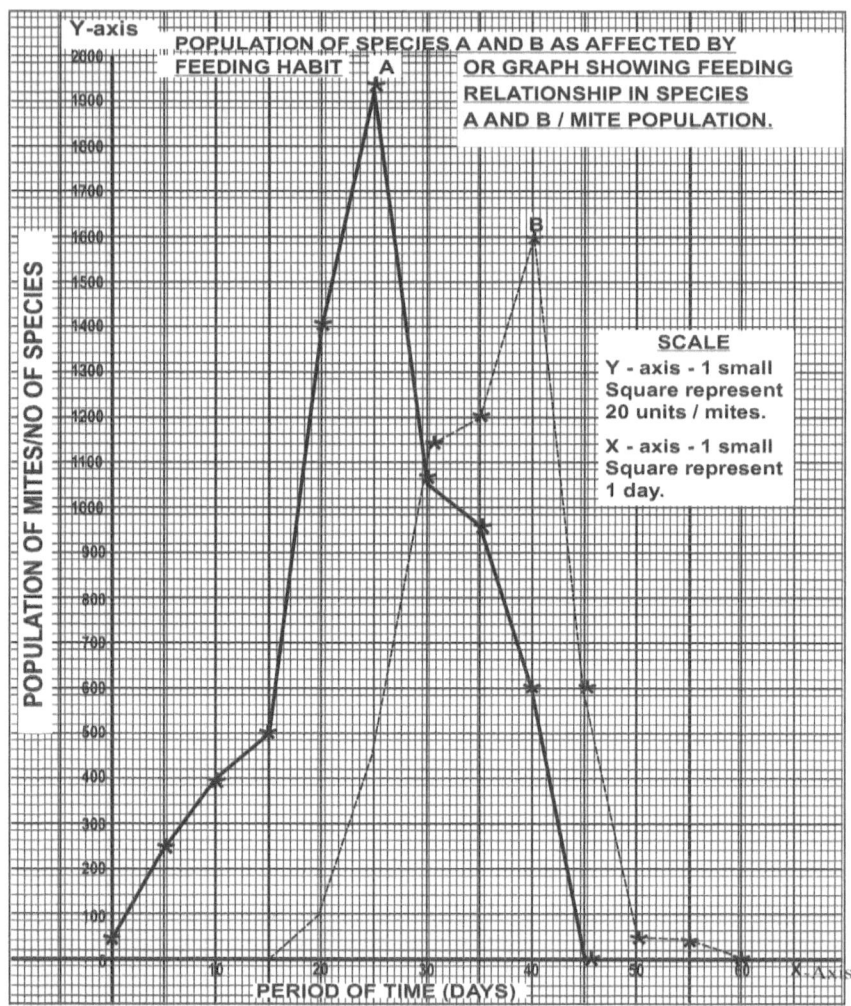

POPULATION OF SPECIES A AND B AS AFFECTED BY FEEDING HABIT OR GRAPH SHOWING FEEDING RELATIONSHIP IN SPECIES A AND B / MITE POPULATION.

Y-axis

POPULATION OF MITES/NO OF SPECIES

2000
1900
1800
1700
1600
1500
1400
1300
1200
1100
1000
900
800
700
600
500
400
300
200
100

A

B

SCALE

Y - axis - 1 small Square represent 20 units / mites.

X - axis - 1 small Square represent 1 day.

10   20   30   40   50   60   X-Axis

PERIOD OF TIME (DAYS)

31B. Description of the two curves

Curve A
> Rose gradually from day 0 to day 15, then rose sharply to a peak at day 25, and then declined sharply to day 30.
> Declined gradually against to day 40, and declined further sharply from day 40 to day 45.

Curve B
> Rose gradually from day 15 to day 25, then rose sharply to a peak at day 40, and declined sharply to day 50.
> No decline at all between day 50 and day 55, followed by a steady decline to zero population at day 60.

31C. Ecological terms

Species A: prey/herbivore/primary consumer
Species B: predator/carnivore/secondary consumer

32A. Count the number of Tridax in each quadrat throw in square

A = Fifteen Tridax Plants
B = Sixteen Tridax Plants
C = Thirteen Tridax Plants
D = Twenty Tridax Plants
E = Fourteen Tridax Plants
> Total number of Tridax plant within five throws of the meter quadrat: 15 + 16 + 13 + 20 + 14 Tridax = 78 Tridax.
> Average number of Tridax per quadrat throw

$$= \frac{78}{5 \times 1m^2} = 15.6m^2 / 15.6 \, per \, m^2$$

- area of abandoned farm land

$$= 80m \times 70m$$
$$= 5600m^2$$

- Estimated population size of Tridax in abandoned farmland

$$= 15.6 \, Tridax \, m^2 \times 5600m^2$$
$$= 87,360 \, Tridax.$$

32B. To know whether it is necessary to control the weeds

To find out whether the weed is increasing in population
To forestall a major spread of the weed on the farmland
To find out the amount of herbicide needed to control the weed
To know the number of laborers to be hired in the weeding of the farmland

32C.

    i.   Parasitic plants:
        Cassytha
        Mistletoe (*Loranthus*).
        Dodder
        Thonningia, aspergillus, penicillium

32D.

    ii.   Parasitic animals:
        Tapeworm (*Taenia*)
        Plasmodium
        Liver fluke (*Fasciola*)
        Ascaris
        Flea
        Hookworm
        Body louse/lice
        Trypanosoma
        Bedbug
        Mite
        Aphid
        Roundworm

33A.

    i.   Microscope/light/electric microscope

ii. Microscope is commonly found in the science laboratory or biology/physics laboratory.
Laboratory science store/preparatory room
Hospitals, diagnostics centers, clinics

iii. Use of object: microscope
For magnifying/enlarging/making bigger the small, tiny, microscopic objects/specimens.
Why/Reasons?
Because the naked eye/ordinary, unaided, natural eyes cannot see the expected details/full details of objects/specimens of study/microorganisms/structures.

33B.

i. Names of labeled parts
I. Eyepiece/ocular/drawtube
II. Body tube
III. Objective/revolving nose piece
IV. Object being studied/slide/specimen
V. Stage
VI. Condenser
VII. Base/base of pillar/foot/stand
VIII. (Moveable) mirror
IX. Coarse adjustment knob
X. Fine adjustment knob
XI. Stage X Fine adjustment knob

ii. Function of the part labeled VII (base/base of pillar): for placing the microscope in position/stand. Firmly act as foot of microscope/instrument/equipment.

34A.

i. Secchi disc
ii. Anemometer
iii. Light meter/photometer/photographic meter
iv. Insect net/Sweep net

v.   Fish trap

34B. Secchi Disc _____ measures transparency or
turbidity/clarity of water.

Anemometer _____ measures speed/velocity of wind.
Photometer/light meter ___ measure light intensity (on land).
Insect/sweep net _____ for catching insects.
Fish trap _____ for trapping/catching fish.

34C. Lower the graduated string into the water

Note the depth of the string when the disc is no longer visible.
Pull the string out gently until the disc becomes just visible.
Note the depth on the string when the disc becomes visible.
Repeat the procedure above.
Take the average of the readings.
The depth at which the disc just cannot be seen.
Measures the turbidity/transparency of the water.

35A.

I.   Amoeba
     Reason
     One-celled organism/unicellular/a cellular
     Irregular in shape
     Presence of pseudopodia/false feet
     Presence of contractile vacuole
     Presence of food vacuole
II. Star fish
     Reason
     Star-shaped
     Head not usually distinct
     Radially symmetrical (at adult stage)
     Possess tough spiny and calcareous exoskeleton

III. Snail (Achatina)
Reason:
Presence of distinct head
Presence of coiled shell
Presence of tentacles (two pairs)
Presence of muscular foot
Knob/eye at the end of second pair of tentacles
IV. Earthworm (Lumbricus)
Reason:
Cylindrical in shape
Elongated body
Pointed at both ends
Segmented body
Presence of clitellum
V. Toad/Frog (*Bufo sp/Rana sp*)
Reason:
Presence of head/bulging eyes/long muscular hindlimbs/
   webbed digits/nostrils/presence of warts/warty skin
VI. Hydra
Reason
Presence of tentacles
It has two-body layers/ectoderm and endoderm/
   diploblastic
Has one opening/hypostome
Presence of bud
Presence of enteron body cavity

35B. Classification

I. Protozoa
II. Echinodermata
III. Mollusca
IV. Annelida
V. Chordata
VI. Coelenterata

36A. Longitudinal/vertical section of a tooth/canine (tooth)

36B.

    i.   Enamel
    ii. Dentine/blood vessel/nerve
    iii. Crown
    iv. Pulp cavity endings
    v.   Gum
    vi. Jaw bone
    vii. Opening for blood vessel nerve/

36C. Tearing/capturing or holding of prey/cutting

36D.

    i.   Nourishes the tooth
        Harbors the blood vessels and nerve endings
    ii. Enamel/I

36E.

    i.   Arch—accidental
        Loop—tented arch
        Whorl—pocket loop
        Double-loop/compound loop
    ii. Crime detection
        Identification
        Appending of signatures

37A. Flame cell

37B.

    A. Nucleus
    B. Cell lumen

C. Flagella/cilia

D. Dust

37C. The beating of the flagella/cilia helps to push water, ammonia in solution, carbon dioxide in solution, and other excretory products toward the tubule dust.

37D. Flatworms/*Platyhelminthes*

38A.

    i.   To show that carbon (iv) oxide/carbon dioxide is necessary for photosynthesis

    ii.

        I.   Soda lime/caustic soda/sodium hydroxide

        II.   Caustic potash/caustic soda/potassium hydroxide

Functions:

        I.   Soda lime _ absorbs the carbon dioxide entering the bell jar

        II.   Caustic potash _ absorbs any carbon dioxide present in the bell jar

        III. Control experiment

        IV. Leaves in B will contain starch/turn blue-black with iodine, while leaves in A will not or remain brown.

39A.

    i.   Cervical vertebra/cervical bone

    ii.   Cervical/neck region/vertebral column/backbone

    iii.

        I.   Neural/neural spine

        II.   Vertebraterial Canal

        III. Centrum

        IV. Transverse process/cervical rib

        V.   Axis—cervical

        Cervical—cervical

Cervical—thoracic

VI.

A. Supports the head/skull
   Permits movement of the head/neck from side to side
   Provides protection for the spinal cord
   Passage of blood vessel
B. Cutin, lignin, cellulose, suberin, pectin
C. Sieve tube (s)
   Companion cell(s)
   Phloem parenchyma
   Phloem fiber

40A.

i. Spirogyra/filamentous alga
ii. Made up of more than one cell joined end to end/filament
   Presence of chloroplast
   Presence of pyrenoid in the chloroplast

40B.

i.

A. Chloroplast
B. Nucleus
C. Pyrenoid
D. Mucilage/gelatinous sheath
E. Cell wall
F. Vacuole
G. Cytoplasmic strand

ii.

A Food manufacture/photosynthesis
B Controls the activity of the organism
C Storage of food/starch/carbohydrate
D Prevention of desiccation/drying

40C. Thallophyta

40D. Rhizopus

40E.

    A. Spore
    B. Sporangium
    C. Sporangiophore
    D. Stolon
    E. Rhizoid

40F.

Diagram I	Diagram II
Filamentous	Branched
Cellular/septate	Non-septate
No sporangia/spores	Sporangia/spores present
No sporangiospores	Sporangiospores present
Slimy	Soft/fluffy/powdery
Photosynthetic/green in color/contain green pigment chlorophyll	Non-photosynthetic/white/black/brown in color/lack green pigment/chlorophyll
Lives in water	Only needs moist condition to grow

40H. Commonly found growing saprophytically on moist organic matter (i.e., bread, garri, etc.).

41A. To demonstrate transpiration in plants.

41B. The bell jar has an opening/it is not airtight/air can enter the setup.

The soil in the pot is not covered by rubber/waterproof/plastic material (to prevent evaporation).

41C. Water droplets/water molecules

41D. Anhydrous copper (II) tetraoxosulphate (VI)/Anhydrous copper sulphate

Blue cobalt chloride paper

42A.

   i.   Trachea/windpipe
   ii. Ribcage/thoracic cage or wall or thorax
   iii. Lungs
   iv. Diaphragm
   v.
      A.   Bronchus
      B.   Thoracic cavity/chest cavity/thoracic chamber

42B.

   i.   I Inhalation/breathing in/inspiration
   ii. II Exhalation/breathing out/expiration

42C. Bell jar I

43A.

   A.   Tooth/incisor
   B.   Esophagus/gullet
   C.   Stomach
   D.   Duodenum

E. Rectum
F. Appendix
G. Large intestine/caecum
H. Small intestine
I. Bile duct
J. Liver

43B.

A. Biting/piercing/gnawing/cutting food into small pieces
C. (Temporary) storage/further digestion of food
H. Digestion/absorption of food
J. Regulation of blood sugar (glucose)/conversion of excess sugar to glycogen
    Regulation of blood protein/deamination of excess amino acid
    Regulation and removal of harmful materials (detoxication)
    Manufacture of fibrinogen/protein metabolism
    Regulation of body heat/heat production
    Lipid metabolism
    Stores fat
    Forms bile
    Stores mineral salts (e.g., iron, copper, potassium)
    Regulates pH balance of the body
    Destruction/breakdown of worn-out/dead red blood cells.

43C.

A. Salivary gland
B. Esophagus/gullet
C. Crop
D. Gizzard
E. Hepatic caeca/mesenteric caeca
F. Midgut/middle gut
G. Small intestine/hindgut
H. Rectum

43D.

    D. Grinding/crushing of food

    E. Production of digestive enzyme/juices/secretion

43E. Muscular

Saclike convoluted/rough interior

43F. Salivary glands, esophagus/gullet, crop/stomach, pancreas/ hepatic caeca, ileum/small intestine, large intestine/rectum/ colon, anus

44A.

    A. Malpighian corpuscle/body/capsule or Bowman's capsule

    B. Branch of renal artery/afferent vessel

    C. Branch of renal vein/efferent vessel

    D. Glomerulus

44B. The difference in relative sizes of B greater than C results in high pressures in glomerulus; this high pressure causes ultrafiltration into capsular space.

44C.

    i. Contractile vacuole

    ii. Nephridia/nephridium

44D. Urine formation in man.

Blood is brought to the kidney by renal arteries as it circulates through the capillaries/glomerulus of each Bowman's capsule.

Water/urea/nitrogenous compounds/mineral salts/sugar/glucose and plasma solutes are filtered into the capsule; ultrafiltration takes place.

The fluid in the capsule/glomerular filtrate flows down the tubule at the proximal convoluted tubule, and Henle's loop some water/sugar/amino acids and salts, which are useful to the body, are reabsorbed into the blood capillaries. Selective reabsorption takes place against concentration gradient/by active transport.

The fluid in the tubule becomes more concentrated as it flows through the distal tubule where more water is reabsorbed, by action of anti-diuretic hormones/ADH, and urine is formed.

44E.

    i.   Nephridia/nephridium
            Malpighian tubules—lenticels
            Flame cells—green glands
            Lung(s)—skin
            Gill(s)—liver
            Lung books
            Stomata/stoma
    ii.   Substances contained in urine
            Urea
            Water
            Traces of uric acid
            Salts/sulphates/phosphates/chlorides
            Traces of hormones

44F. Osmoregulation in man

The kidneys carry out osmoregulation in man when the osmotic concentration of blood is higher than that of the cell content by means of sugar/mineral salts/amino acids/other plasma solutes.

Kidneys extract these substances from the blood/body cells absorb more water.

When these substances are present in small amounts in the blood/higher osmotic pressure in body cells, thus decreasing osmotic concentration of the blood.

The kidney extracts more water than usual.

By the action of ADH (anti-diuretic hormone).

And more concentrated urine is produced, thus keeping the osmotic concentrations of blood and cell fairly constant/ avoiding any damage.

Water can also be lost through sweating.

Water can be lost through breathing out moist air.

45A.

A. Sebaceous gland
B. Sweat pore
C. Cornified layer
D. Granular layer
E. Malpighian layer
F. Erector muscle
G. Sweet duct
H. Nerve fiber
I. Pressure sense
J. Sweat gland
K. Blood vessel
L. Fatty cells

M. Hair shaft
N. Pain receptor
O. Touch
P. Heat receptor
Q. Cold receptor
R. Hair follicle
S. Blood capillary to the hair

45B. A Sebaceous gland secrets oily substance [sebum] which lubricates the hair. The sebum contains lysozyme which kills bacteria and viruses.

K. Blood vessel supplies food and oxygen to the tissues of the skin.
L. Fatty cells act as insulating layers. They also act as energy reserve and protect skin from damage.

45C. C, D, and E: cornified layer, granular layer, and Malpighian layer.

45D. Description of skin

Two major layers:
The upper/outer epidermis and a lower/inner dermis

Epidermis consists of thin dead cells, is horny, and has cornified layer, below which is the granular layer of living cells that divide continuously to produce new cells to replace those lost from the surface.

The dermis is a thick/dense active layer which contains connective tissues, glands, nerves, blood vessels, hair follicle/hair, erector muscle, receptor cells, and adipose tissues.

45E. Regulation of body temperature.

There is excess heat in the body.

Blood flow in the dermal capillaries is stimulated.

The blood vessels under the skin epidermis dilate.

More blood is brought to the surface.

The sweat glands become more active/extracts large quantities of sweet from blood, and sweat is produced.

The sweat picks up the latent heat of vaporization from the skin.

Cooling the skin down.

Heat is also lost by radiation, allowing air to flow through the follicles.

Heat is lost through convection.

The hair lies inclined/flat.

45F.

i. Sensory function: possession/presence of sense organs/ receptor cells to touch, to feel pain, heat, or cold to receive stimuli.
   And nerve fibers which transmit impulses to the central nervous system for interpretation.
ii. Excretory function: sweat glands extract water; sodium chloride and urea. Excretory products form the blood capillaries.
   Which are secreted to the sweat duct and finally to the sweat pore
iii. Thermoregulation function: when body temperature rises.
   Vasodilation occurs and increases blood supply to the surface of the skin.
   Heat is lost by radiation or convection.
   When body temperature reduces/fails or is low, vasoconstriction decreases blood supply to the surface of the skin.
   Heat is conserved in the body

45G.

The skin	A leaf
Sweat pore/pore present	Stomata/pores present
Epidermal layer present	Epidermal layer present
Cornified layer/protective outer layer present	Cuticle/protective outer layer present
Hair on epidermal layer	Hair on epidermal layer
Blood vessels/capillaries/conducting tissues present	Vascular bundles/conducting tissues present
Stratification shown	Stratification shown

45H. For protection

Excretion and osmoregulation.
Maintenance of a suitable constant body temperature.
The skin manufactures and stores vitamin D.
Storage of reserved food.
The mammary glands in mammals are modification of the skin; they help to produce milk for feeding the young.

46A.

A. Nasute soldier   D. Winged reproduction
B. Worker   E. Queen
C. Mandibulate soldier F. King

46B.

1. Worker(s)
2. Soldier(s)
3. Queen
4. King
5. Reproductive(s)

46C. Worker: build/repairs the nest

Gather food
Take care of the nymphs
Clean the nest
Feed the nymphs/queen/king/soldier
Collect eggs
May kill/starve the queen when too aged
Soldier: they defend/protect the nest from enemies.
Protect the workers during food gathering.
Queen: lay eggs/produces eggs/reproduction.
King: reproduction/fertilizes the eggs/mating with the queen.
Reproductive: they are future queens and kings of new colonies.
They help to form new colonies.

46D. Destructive to crops/wood/furniture.

Improves aeration of soil
Edible/source of protein/source of revenue earning
Helps to speed up decay processes

46E. They live together in nests or colonies.

There are distinct castes or individuals.
There is a clear division of labor.

	Gland	Location
i.	Adrenal	F
ii.	Pancreas	C
iii.	Thyroid	B
iv.	Pituitary	A
v.	Ovary	D
vi.	Testes	E

47A.

    A. Pituitary gland
    B. Thyroid gland
    C. Pancreas
    D. Ovary
    E. Testes
    F. Adrenal gland

47B.

Secretion	*Gland*
i. Thyroxin	Thyroid
ii. Adrenaline	Adrenal
iii. Insulin	Pancreas
iv. Estrogen	Ovary
v. Testosterone	Testes

47D.

(C)	Oversecretion	Deficiency
(i) Thyroxin	- Bulging eyeballs - Hyperthyroidism/ increased - Metabolic rate/ restlessness/ - Overactivity/ anxiety - Accelerated heartbeat - Loss of weight	- Hypothyroidism/ low metabolic rate/sluggishness/ cretinism/goiter - Increase in weight - Mental activity slows down - Cretinism in infants
(ii) Adrenaline	- Hypersensitivity/ overanxiety - Or excitement/ masculine - Sensation in female	- Lethargy/lassitude/ indifference/fall in blood pressure - Lack of energy/ weakness/tiredness

(iii) Insulin	- Fall in blood sugar level - Hypoglycemia - Incessant hunger	- Diabetes mellitus/ high blood pressure - Excess sugar in blood/urine - Hyperglycemia - Low/less appetite
(iv) Estrogen	- Abnormal urge for sex/high urge for sex - Early maturity to secondary sexual character in females	- Poor development of the reproductive system/ low urge for sex - Delayed secondary sexual maturity
(v) Testosterone	- Abnormal urge for sex	- Poor development of reproductive system - Decline of male secondary sex characteristics

47F. Secretes pancreatic juice

Which contains three digestive enzymes, namely amylase, trypsin, lipase

Amylase converts starch to maltose

Trypsin converts proteins to peptones, polypeptides, and amino acids

Lipase acts on emulsified fats, changing them to fatty acids and glycerol

Chymotrypsin converts casein to polypeptides

Carboxypeptidase converts polypeptide to amino acids

48A. Fingerprints/finger impressions

48B.

1. Arch
2. Loop
3. Whorl
4. Double loop/compound loop

48C.

A. Arch
B. Tented arch
C. Loop
D. Plain whorl
E. Accidental
F. Central pocket loop
G. Double loops

48D. For crime detection; for identification.

Appending of signatures.

Reasons:

The contouring/ridges of the fingerprints/impressions differ or peculiar to individuals.

No two individuals have the same contouring/ridges.

Uniqueness of individual, hence crimes committed by individual persons are easily detected.

48E. Heredity or genetics/variation/morphological variation.

49A. Axile (placentation)

49B. Tomato/canna lily/okro/citrus/grapes/melon

49C.

   A. Seed
   B. Placenta

49D. Placentation is the arrangement of ovules in the placenta of the ovary

49E.

   1. Marginal placentation (e.g., *Delonix regia* (flamboyant), crotalaria, cassia)
   7. Parietal placentation (e.g., Paw-paw)
   8. Axile placentation (e.g., tomato, orange/citrus/grape)
   9. Free central placentation (e.g., waterleaf, coconut)
   10. Superficial (e.g., waterlily)
   11. Basal (e.g., sunflower)

50A.

   A. Organic matter/humus
   B. Water with dissolve particles of clay
   C. Clay
   D. Silt
   E. Fine sand
   F. Coarse and gravel

50B. Experiment to show that soil contains different sizes of particles/ separation of soil particles from humus by sedimentation to prepare soil profile.

50C. Improves soil fertility.

Provides plant nutrients, especially nitrogen/phosphorus/ sulfur.

Release of organic acid during decomposition helps to dissolve.

Minerals in rocks.

Organic acid/matter increases cation exchange in soil.

Lower soil acidity pH.

Decomposition of humus/organic matter releases carbon(iv) oxide and ammonia to the atmosphere.

Reduces soil erosion by binding soil particles together.

Allows burrowing of earthworms, ants, and rodents into the soil; for better air and water infiltration.

Reduces evaporation by formation of surface mulch.

50D. Provides anchorage

Provides medium for both microbial and microbial activities.

Improves soil fertility.

For better plant growth.

Provides plant nutrients.

Provides raw materials for photosynthesis (e.g., water).

Contains soil water/dissolved inorganic minerals for healthy growth of plants.

51A.

A. Ovule
B. Placenta
C. Placenta

51B.

I. Marginal placentation
    Reason: Ovary is formed from a single carpel joined along its edges.
II. Parietal placentation
    Reason: the carpels of a syncarpous ovary are joined to one another by their edges to form a single chamber or loculus with the placentae on the walls.
III. Axile placentation
    Reason: the carpels of a syncarpous ovary do not meet at the edges but are pushed in so that the walls meet in the middle of the ovary, and the ovules are borne on the central axis so formed. Has two or more chambers or loculi.
IV. Free central placentation
    Reason: the syncarpous ovary contains one large loculus, and the ovules are borne on a knob which projects from the base of the ovary.

51C.

I. Marginal placentation [e.g. cassia, crotalaria, *Delonix regia* (flamboyant)]
II. Parietal placentation (e.g. Paw-paw)

III. Axile placentation (e.g., tomato, canna lily, orange)
IV. Free central placentation (e.g., waterleaf, coconut).

52A.

    I.   Tadpole with limbs
    II. Bony fish/tilapia

52B.

    I.   Streams/ponds
    II. Pond/lake/river/freshwater/stream

52C.

    I.   Phylum: Chordata class: Amphibia
    II. Phylum: Chordata class: Pisces/osteichthyes

52D. Presence of streamlined body

Presence of scales
Presence of operculum
Presence of homocercal tail
Possession of gills

52E.

A.	Eye	I.	Operculum or gill cover
B.	Nostril		
C.	Mouth	J.	Tail fin/caudal fin
D.	Forelimb with four digits	K.	Scale
		L.	Anal fin
E.	Tailfin	M.	Pelvic fin
F.	Tail muscle	N.	Pectoral fin
G.	Dorsal fin	O.	Operculum or gill cover
H.	Tail (fin)/caudal fin		

P. Mouth   R. Eye: four digits
Q. Nostril   S. Lateral line

52F. Functions of parts labeled

    G. Prevents body from rolling sideways/for swimming/ weapon of defense.

    I. Allows easy movement.
      Protects the body from mechanical injuries and microbial infections.
      Prevents loss or gain of water through the skin.

    M. Protects the gills/exit for water to flow over gills during feeding/respiration/breathing/gaseous exchange.

52G. Frog/toad

52H. Tail

Tail/caudal fin
Nostrils
Mouth
Eyes
Operculum/gill cover/operculum fold
Streamlined body

    I.

Organism I (Tadpole with limbs)	Organism II (Tilapia/ bony fish)
1. Long tail	Short tail
2. Limbs present	Limbs absent
3. Scales absent	Scales present
4. Lateral line absent	Lateral line present
5. Has only one fin/caudal fin	Has many fins

52I.

Features	Functions
Streamlined body	For rapid movement with least resistance
Caudal fin	For steering
Muscular tail	Forward propulsive force
Median/unpaired fins/dorsal fin	For stability in water
Paired fins	Steering in water
-Overlapping scales	Rapid movement/protection
Lateral line	Detection of vibrations
Operculum/gill cover	Protects the gills/ for breathing
Mouth	For inflow of water for respiration
Dark gray dorsal/ silver ventral surface	For camouflage against predator

52K. Proteins

Fats and oils
Vitamins (various)

53A.

Identification	Phylum
A—Earthworm	Annelida
B Bird/hen/chicken	Chordata
C Spirogyra	Thallophyta/Algae/
D Crab	Chlorophyta
E Periwinkle	Arthropoda
	Mollusca

53B.

    i.   Cell membrane → Gills   →   Skin   →   Lungs
        (Spirogyra) → (Periwinkle) → (Earthworm) → (Bird) Crab

   ii. Most primitive respiration surface is the *membrane.*
        Reasons: It is very simple—not complex—was the first to
        evolve, and very thin.

53C.

    i.

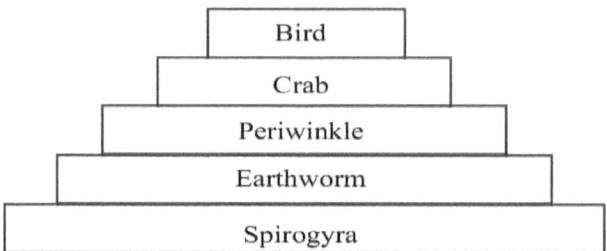

   ii. Organisms with greatest amount of captured energy is
        spirogyra.

  iii. Least amount of captured energy bird.
       5.   Producer spirogyra
       6.   Primary consumer earthworm
       7.   Secondary consumer periwinkle
       8.   Tertiary consumer crab/bird

53D. The transfer of energy between tropic level is not 100%.

Successive levels have less of useful energy and support fewer
organisms.

Primary producers/plants have the highest amount of energy.

When herbivores feed on plants, the energy level is reduced.

When carnivores consume the herbivores, the energy level is further reduced.

53E.

1. Pyramid of numbers
4. Pyramid of biomass
5. Pyramid of energy

54A. Larva/larval stage of *culex*, mosquito/mosquito larva.

54B.

A. Siphon/siphon tube/breathing tube/breathing trumpet
B. Hair
C. Antenna
D. Mouth brush/part

54A. Stagnant water/water in a container/ponds/lakes/puddles/pool/standing water/still water

54B.

A. Used for breathing/taking in air at water surface
B. Used for sweeping small organisms in water into the mouth/used for feeding.

55A.

i. Tick/dog tick (*Boophilus*)
ii. Ectoparasite/external parasite
iii. Cow/cattle, goats, sheep, dogs, etc.

55B.

i. Symbiosis/mutualism

ii. Lichen/alga(e)/fungus/fungi
iii. Rocks surface of leaves
Tree barks/trunks
Surface of soil/wall
iv. Two plants organisms
Nitrogen-fixing bacteria/rhizobium/*Rhizobium legumino-*
sarum in root nodules of legumes
Lichen/Alga cells/fungi hyphae
In Animals:
Sea anemone on hermit crab
Ox pecker on buffalo
Zoochlorella in hydra
Trichonympha in termite
v. Parasitism
Commensalism
Saprophytism

56A.

I. Yeast/yeast cells/yeast multiplying by budding/saccharomyces
II. Yeast cell
III. Mushroom

56B. Thallophyta/basidiomycota/ascomycota

56C.

A. Bud
B. Yeast cell
C. Nucleus
D. Nuclear vacuole
E. Chromatin thread
F. Glycogen
G. Volutin granule
H. Pileus
I. Gill(s)

J. Stipe
K. Hyphae

56D. They are nongreen plants

The cell wall is composed of cellulose and chitin.
They are simple multicellular plants.
They are either saprophytic or parasitic.
Reproduction is by means of spores.
Carbohydrates are stored in form of glycogen as in animals.
The vegetative body (hypha) is collectively known as mycelium.
The body is not differentiated into root, stem, and leaves.

56E. Occurs naturally in the juice of fallen overripe fruits.

Sugar-containing solutions such as palm wine.

56F. Production of alcoholic drinks and industrial spirit.

Bakers use yeast to make bread rise.
Important source of vitamin B.

56G.

i.   Terrestrial/decaying matter in the soil.
iii. Umbrella-shaped structure which consist of a stalk or stipe and an expanded cap or pileus.
     Presence of interwoven hyphae.
     Radiating out from the stipe on the underside of the pileus are a number of delicate gills.
     The pileus is the reproductive part.
     The gills are made of interwoven hyphae which bend outward toward the surface.
     The end cells of the hyphae, which are called basidia, bear spores.

56H. Mucor

Rhizopus
Aspergillus
Toadstool
Penicillium
Fusarium

57A.

    i.   Pooter
    ii.
        I.   Glass mouth piece/air outlet tube
        II.   Glass/plastic/rubber tube
        III. Net with fine mesh
        IV. Specimen bottle
        V.   Insect
        VI. Air inlet tube
    iii. For collecting small insects from hard surfaces/tree trunks/ leaves/walls etc.

58A.

    i.   Potter

58B.

    ii. Catching/collecting/sucking small animals/insects from leaves/plant surfaces, walls, rocks, or crevices.

58C. To prevent dirt/caught specimen from entering the suction tube

58D.

    B. Collecting inlet tube
    C. Suction tube/tube for sucking air

59A.

    i.   The setup is on hydrotropism/movement of the root toward source of water

    ii.

        I.   Water

        II.   Shoot (of seeding)

        III. Sawdust/soil

        IV. Trough/container

        V.   Seed/grain

        VI. Root/radicle

        VII. Porous pot/unglazed clay pot/epicotyl/stem

        VIII. Holes/perforations

59B. Precautions

Sawdust/soil used must be fairly dry in order to create moisture gradient.

Seed should be planted a few centimeters away from the porous pot to enable response to moisture gradient.

Mark the side of the seedling facing source of water to indicate the direction of root curvature.

Carefully remove the seedling at the end of the experiment to avoid breakage of radicle/root.

59C.

    i.   The part labeled VI will grow straight in to the soil/downward in response to the stimulus of gravity.

    ii.   The part labeled I (water) is more important because without I, II cannot germinate; or even if it germinates, II will not respond "hydrotropically."

59D. That plant root grows toward the source of water, in defiance of gravity, to show that the root of the plant is positively hydrotopic.

60A.

    I.   Presence of brick red (precipitate)/brownish red/red/ orange (precipitate).

60B. Simple/reducing sugar/any correctly named reducing sugar (e.g., glucose).

60C. It hydrolyzed the food substance to simple/reducing sugar/ hydrolysis/hydrolyzed the food substance.

60D. To make the mixture alkaline/to neutralize the acid.

60E. There was no reducing sugar/reducing sugar absent.

60F. Carbohydrate/starch/complex sugar/non-reducing sugar/any example of non-reducing sugar (e.g., sucrose, lactose, maltose).

    II.
- A. Starch: iodine solution
- B. Reducing sugar: Fehling's solution or Benedict's Solution
- C. Protein: sodium hydroxide, copper sulphate, or Millon's reagent
- D. Fat: Sudan III solution
- E. Complex sugar-dilute hydrochloric acid, dilute sodium hydroxide, Benedict's solution, or Fehling's solution.

61A.

A. Olfactory lobe
B. Anterior cerebral hemisphere
C. Posterior cerebral hemisphere
D. Pineal body
E. Vermis
F. Flocculus
G. Cerebellum
H. Spinal cord
I. First spinal nerve
J. Spinal cord
K. Medulla oblongata
L. Optic lobe

61B.

i. Thalamus
   Helps in experiencing sensation of pains/touch/anger.
   Seats of consciousness or awareness.
   Receives impulses from midbrain, hindbrain, and spinal cord.
   Sends/passes impulses to cerebral cortex and vice versa.

ii. Medulla oblongata
    Controls heartbeat/involuntary actions.
    Controls breathing rate/respiration.
    Controls blood circulation, swallowing, vomiting, and salivation.
    Controls digestive movement.
    Controls laughing.

iii. Hypothalamus
     Controls sleep/alertness.
     Controls appetite/feeding.
     Controls body temperature.
     Controls amount of water in blood/osmoregulation.
     Controls the secretion of hormones from the pituitary gland.

iv. Cerebellum
    Body balance or positioning of the body/posture.
    Coordination of various muscle actions in involuntary responses.

62A.

    III. Cervical vertebra
        Presence of Vertebrarterial Canal
        Short transverse process/cervical rib
        Short neural spine and neural spine points backward
        Large neural canal
    IV. Thoracic vertebra
        Long neural spine
        Large/prominent centrum
        Short/prominent transverse process
        Facets for articulation with ribs
    V. Lumbar vertebra
        Large centrum
        Prominent/large transverse process
        Presence of anapophysis and metapophysis
        Short neutral spine
        Prezygapophysis curves inward

62B.

    A. Neutral spine: attachment of muscles.
    B. Transverse process: attachment of muscles.
    C. Neutral canal: passage of spinal cord/encloses the spinal cord.
    D. Capitular demi-facet/articular facet: articulates with the next vertebra.
    E. Centrum: supports the weight of the vertebral column/ point of articulation for adjoining vertebra.
    F. Prezygapophysis/articular facet: articulates with the next vertebra/articulates with the ribs.
    G. Vertebrarterial Canal: passage for blood vessels.

63A.

A. Dendrites
B. Nucleus
C. Cytoplasm
D. Neurilemma
E. Schwann cell
F. Myelin sheath

G. Node of Ranvier
H. Axon
I. Terminal dendrite
J. Axon terminal/bouton
K. Cell body
L. Axon

63B. A neuron is made up of an axon and cell body.

The cell body consists of nucleus, cytoplasm, dendrons/dendrites.

The axon is enclosed within a myelin sheathe/fatty sheathe with Schwann cells located on it.

Sheathe is interrupted at intervals by constrictions/nodes of Ranvier.

Nuclei are located at intervals within the myelin sheathe.

63C.

1. Sensory/afferent neuron receives impulses from organs and conveys them toward the brain/spinal cord/central nervous system (CNS).
   (Carries impulses to the CNS)
2. Relay/intermediate/connective association neuron; receives impulses from sensory neuron and conveys them to motor neuron.
3. Motor neuron/efferent neuron carries responses from the brain/spinal cord/CNS to the target organ/effector (takes impulses away from the CNS).

63D. The transmission of impulses is by change in electrical potential and ionic charge—at rest/resting potential, the concentration of ions on either side of the membrane is polarized.

The outside of the nerve is positively charged.

The inside is negatively charged.

Chemically, the nerve fiber has potassium ions concentrated inside the fiber.

Sodium ions are concentrated more on the outside.

When the fiber is stimulated/depolarized.

There is a change in permeability of the cell membrane.

Sodium ions enter the nerve by diffusion.

Potassium ion moves to the outside.

This change in permeability along the nerve/change in electrical potential and ionic charge along the fiber.

Causes the movement of electrical current along the fiber.

In this way, the impulses are transmitted, and the original resting stage of the nerve is restored as soon as the impulses pass.

63E. Differences

I. (Cervical)	II. (Thoracic)
1. Short neural spine	Long neural spine
2. Reduced/short transverse process	Prominent/longer transverse process
3. Large neural canal	Small neural canal
4. Vertebrarterial Canal present	Vertebrarterial Canal absent Tubercular/capitular facets present
5. Tubercular/capitular facets absent	

64A.

A Pelvic girdle
Observable features:
Presence of ilium, ischium, and pubis which fuse together to form innominate bone
Pubis fusing together/pubic symphysis
Presence of concave articulating surface called acetabulum
Presence of large holes/obturator foramen
Presence of facet for articulation with sacrum.
B Femur
Observable features:
The structure has a long shaft with a round head at one end
The head has a short neck
Presence of the great trochanter/lesser trochanter
Presence of patellar groove and condyle

64B.

A. Ilium
B. Sacrum
C. Acetabulum
D. Obturator foramen

E. Ischium
F. Pubic symphysis
G. Pubis
H. Sacroiliac joint

I.   Greater trochanter   M. Condyle
J.   Third trochanter   N. Lesser trochanter
K. Shaft   O. Neck
L.   Patellar groove   P.   Head

64C. Functions of A (pelvic girdle):

Has surface for muscle attachment (for movement).

Protects some reproductive organs.

Attached to the sacrum by ilium/supports the sacrum.

Forms ball-and-socket joint with femur (i.e., for movement of the lower limb).

Has the obturator foramen as a passage for nerves and blood vessels.

Functions of B (femur):

Supports the upper leg/thigh.

Formation of red blood cells/erythrocytes in the cells of the bone marrow.

Provides articulating surfaces for the tibia and fibula.

Forms ball-and-socket joint with the pelvic girdle.

The patellar groove articulates with the patellar/knee cap.

64D. How bone A/pelvic girdle is adapted to articulate with bone B/femur:

The concave articulating surface/acetabulum is deep and admits the entire head of bone B/femur to form the ball-and-socket joint.

The short neck suspending the head of femur assists in this union.

64E. Nutrients contained in A and B:

Calcium/phosphorus/calcium phosphate

Functions:

Formations of strong, healthy teeth/bones.

Calcium is for blood clotting/coagulation.

Calcium is for proper functioning of the heart and nervous system.

Calcium is for normal contraction of muscles.

Phosphorus regulates metabolism of proteins, fats, and carbohydrates.

Phosphorus is a components of DNA/RNA/nucleic acid.

65A.

    I.   Thermometer
    II.   Hygrometer
    III. Anemometer
    IV. Wind vane

V. Barometer
VI. Photometer/light meter
VII. Rain gauge
VIII. Insect net/sweep net

65B.

I. For measuring temperature.
II. For measuring relative humidity.
III. For measuring wind speed.
IV. For finding direction of wind.
V. For measuring atmospheric pressure.
VI. For measuring light intensity.
VII. For measuring amount of rainfall.
VIII. For collecting insects.

65C.

I. Thermometer

Instrument placed appropriately in environment.
Allow to stabilize.
Take reading of the thermometer.
Record your observation.
Repeat/confirm after a given interval of time.

VII. Rain gauge

Mount instrument on raised platform/partly bury rain gauge and fix an elevated funnel devoid of any shade, like tree or house.
Check the bottle to see that it is empty and clean and graduated/calibrated (use measuring cylinder).
Take reading of volume of rainwater in bottle after rainfall.
Record your observation.

65D. Fault with apparatus

Presence of formalin in apparatus

Absence of muslin cloth on the suction section of the apparatus.

Delivery tubes for inlet and outlet are of the same lengths. Collecting tube should be longer.

66A.

i.
- I. Larva/larval stage of housefly/maggot
- II. Imago/adult housefly
- III. Pupa stage of housefly

ii.
- A. Mandibular sclerite/hook/mandible
- B. Anterior spiracle
- C. Spiniferous pads/spiny pads
- D. Posterior spiracle

66B.

- A. For feeding/for locomotion.
- B. For gaseous exchange /for breathing/for respiration.
- C. For movement/for locomotion.

66C. Transmission of harmful disease.

66D. Hairy body/legs/appendages/abdomen

66E. When the organism perches or lands on contaminated stool or decaying substances.

Disease-causing organisms/bacteria/viruses adhere to the hairy body and legs.

When it lands/perches on food or water.

It deposits some of the disease-causing organisms/cholera/ germs on the food or water.

When the food or water is taken in, it may cause the disease.

66F.

    I.   White/cream/butter color
    II. Gray/brown/green

67A.

Seeds must be viable/germinating.
All joints of the apparatus must be airtight.
Measuring tube must dip into water in the beaker.

67B. 70–10 cm³ = 60 cm³/cc. (Read from the setups.)

67C. 5 cm³/cc (per hour).

67D. Oxygen in flask is used by respiring seeds.

Carbon(iv)oxide is produced by respiring/germinating seeds.

The carbon(iv)oxide is absorbed by potassium hydroxide.

This causes reduced pressure within the flask.

Because the system is airtight, a vacuum is created within the apparatus is replaced by water from the beaker, which is manifested as rise in water level within the measuring tube.

The amount/volume of carbon dioxide absorbed is equivalent to the amount of water that moves into the tube.

67E. Warmth/adequate temperature

Moisture/water
Air/oxygen

68A.

I.   Littoral zone
II. Benthic zone

68B. High light intensity/bright light/well-illuminated.

Wide variations in temperature.
Sufficient oxygen/high oxygen supply.

68C. Plants: Water lettuce, duckweed, ceratophyllum, azolla, water lily, spirogyra, chlamydomonas, volvox, pandorina, eudorina.

Animals: Birds, cyclops, daphnia/water fleas, mosquito larva, mosquito, tadpoles, pond skaters, dragonfly, dragonfly larva, euglena, paramecium, amoeba.

68D. Zone I/littoral zone.

68E. In Zone I, there are many green plants and chlorophyll-possessing organisms that are capable of carrying out the process of photosynthesis.

68F.

Plant in Zone I	Plants in Tropical Rainforest
Poor root system	Well-developed root system
Can float in water because they have buoyancy adaptations	Very heavy plants that cannot float in water but are properly established in the soil
Thin cuticle/no cuticle which allows easy absorption of water	Have thick cuticle which prevents water loss
Tissue spongy/herbaceous	Tissue woody
No canopies	Canopies prominent

68A. Bacteria, fungi, algae, insects, fishes, periwinkle/snails/bivalves.

69A.

    i.
- I. Orchid (epiphyte)
- II. Mangrove/rhizophora/plant with stilt roots
- III. Climbing/twining plant/lianas/climber/twiner
- IV. Frog/tree frog

    ii.
- A. Orchid/epiphyte, or shoot of orchid/epiphyte
- B. Aerial root
- C. Stilt roots
- E. Adhesive/suction pads/digits/toes

69B.

- I. Tropical rainforest
- II. Swamp/estuarine/mangrove forest/swamp
- III. Tropical rainforest
- IV. Tropical rainforest/swamp

69C. Heavy rainfall/high humidity/abundant water supply

Canopies in strata/layers
Tall trees
Dense foliage at crown
Low-light intensity
Little or no under growth
Presence of epiphytes/climbers/buttress roots.

69D.

B. Aerial roots—hairy/spongy for absorbing water from tree trunk/humid atmosphere/absorbent.
C. Stilt roots—highly branched/widespread/multiple/arched roots: extra support for plants/mangrove in soft mud.
D. Climbing   stem—flexible/climbing/twinning/twisting stem: for support or to facilitate reaching the top of canopies of rainforests for sunshine.
E. Adhesive pad/digit/toe—sticky adhesive pad/flexible/ elongated digit: for grasping a strong hold for climbing trees.

70A.

I.   Earthworm
II. Snail/Achatina

70B.

I.   Annelida
II. Mollusca

70C.

A. Prostomium
B. Mouth
C. Chaetae
D. Female reproductive opening
E. Clitellum
F. Opening of sperm duct
G. Moist skin
H. Anus
I. Shell
J. Occuliferous tentacle
K. Eye
L. Tactile tentacle
M. Mouth
N. Genital pore
O. Collar
P. Foot
Q. Respiratory pore
R. Anus

70D.

C/Chaetae: chaetae function as a holdfast when the worm is burrowing or moving on the surface of ground.

E/Clitellum: secretes materials for making cocoons containing eggs.

70E.

Earthworm burrows into the soil, thereby aerating the soil.

They loosen the soil particles and make it easier for plant roots and water to penetrate.

Drag humus into their burrows, thereby increasing the humus content of the soil/improving soil fertility.

The burrow activities turn/mix the top and subsoils together for better quality of soil.

Their fecal matter/worm cast adds to soil nutrients used as bait in fishing.

70F.

    I. Moist soil rich in humus (terrestrial/land).
    II. Land (terrestrial)

70G. Male sex organ = F

Female sex organ = D
Hermaphrodite

70H. The organism is active/come out in the night.

    i. Color of shell (serves as camouflage)
    Slimy secretion
    Hard shell

71A.

    I. Prawn (*Palaemon*)
    II. Cockroach

71B.

I Aquatic freshwater/marine/stream/river/sea
II Terrestrial dark cupboard/dirty toilet

71C.

A. Rostrum	I. Uropods
B. Eye	J. Swimmerets
C. Antennules	K. Telson
D. Antenna	L. Abdominal segment
E. Third maxilliped	M. Carapace
F. Claw	N. Antenna
G. Chela	O. Compound eye
H. Third walking leg	P. Prothorax

Q. Mesothorax
R. Metathorax
S. Jointed leg
T. Abdomen
U. Anal cercus

V. Style
W. Spiracle
X. Membranous wings
Y. Elytron

71D.

	Phylum	Class
I	Arthropoda	Crustacea
II	Arthropoda	Insecta

71E. Economic importance of insects

Act as carriers/vectors/transmitters of diseases (e.g., Anopheles mosquito/Tsetse fly).

Act as pests, destructive to live plants/stored food (e.g., locust/grasshopper/weevil/crickets).

As food/source of protein (e.g., termites, honey, larval forms of some beetles).

Used in medicine (e.g., bees, dung beetle, caterpillar of some beetles).

Phylum class

I. Arthropoda crustacea
II. Arthropoda insecta
   Used in biological control (e.g., ladybird used in control of aphids on plantations; praying mantis control of some insects).

   Serve as pollinating agents of flowering plants (e.g., bees, butterflies, ants, moths, wasps, beetles).

Soil insects for aeration/loosening of soil (e.g., termites, crickets, dung beetles).

Insects as industrial materials (bee wax for candles; silk material from silk worms).

Termitarium: used in building lawn tennis courts.

Revenue yielding (e.g., termites, larval of beetles etc.).

Destroy furniture, books, clothes, wood, etc. (e.g., cockroaches, termites).

71F. Rostrum: for offence and defense.

Swimmerets: for swimming.

Chela/pincers: for holding food/offence/defense.

Telson/uropod/tail plate: for swimming/steering/darting.

Antennule: for smelling, hearing, balancing, and also sensitive to touch.

71G.

I.   Cephalothorax and abdomen
II. Head, thorax, and abdomen

71H. Labrum (upper lip)

Mandibles
Maxillae
Labium (two maxillae)

71I.

    I. Water fleas, sacculina, barnacles, shrimps, woodlice, cray-fish, lobsters, crabs.

    II. Locusts, grasshoppers, crickets, dragonfly, butterfly, moths, housefly, mosquito, honeybee, ants, wasps, beetles, bug.

71J. Compound eyes, antennae, exoskeleton, jointed appendages, segmented body.

71K. Male cockroach because of the presence of anal style.

71L. The body is covered with exoskeleton made of chitin which helps to protect the internal organs, provide places for attachment of muscles, and to prevent water from entering and leaving the body.

Jointed appendages for easy locomotion.
Compound eyes ensure keen sense of sight.
Tracheal system of air tubes which ensure direct delivery of oxygen to the tissues.

72A.

    A. Scapula/shoulder blade
    B. Tendons
    C. Humerus
    D. Triceps (muscle)
    E. Ulna
    F. Radius
    G. Biceps (Muscle)

72B. Ball-and-socket joint

72C. Hinge joint: elbow/knee/fingers.

Rotating/pivot joint: in the neck/atlas/skull.
Gliding joint in the wrist/ankle/between vertebrae.
Suture joint in the skull.

72A.

    i.   Synovial fluid
    ii. For movement

73A.

    I.   Digestive system of planaria
    II. Digestive system of earthworm.

73B.

A. Eye	F. Mouth
B. Mouth	G. Pharynx
C. Pharynx	H. Esophagus
D. Diverticular (intestinal branches)	I. Crop
E. Intestine	J. Gizzard
	K. Intestine

73C. Three main branches: crustaceans, rotifers, nematodes, insects.

73D.

Crop: It functions as a temporary storage chamber.

Gizzard: for grinding food into small particles by its churning action.

73A. Planaria.

74A.

    I.   A Typical animal cell.
    II. A Typical plant cell

74B. Similarities

They both possess:
Cytoplasm: nuclear membrane
Cell membrane: nucleus
Endoplasmic reticulum: chromosomes
Golgi apparatus: nucleoplasm
Mitochondria: nucleolus
Ribosomes: they both carry out mitosis in somatic cells and
    meiosis in reproductive cells.

Differences:

I (Animal Cell)	II (Plant Cell)
Chloroplast is absent	Chloroplast is present
-Vacuoles usually absent but, if present, are small and temporary	Large permanent central vacuole is present
Food is stored as glycogen and fat	Food is stored as starch granules
Centriole is present	Centriole is absent
Cell has living cell membrane, hence it can change its shape	Cell has a dead cellulose cell wall, hence definite in shape
Cytoplasm spreads all over the cell	Cytoplasm pushed to the cell wall

74C.

A. Cell membrane
B. Cytoplasm
C. Centriole
D. Nucleus
E. Ribosomes
F. Lysosome
G. Mitochondrion
H. Golgi apparatus
I. Cell membrane
J. Golgi apparatus

K. Chloroplast
L. Nucleolus
M. Endoplasmic reticulum
N. Vacuole
O. Mitochondrion
P. Cytoplasm
Q. Cell wall/plasma membrane

74D. The cell theory:

New cells arise from preexisting cells-by-cell division.

The cell is the structural and functioning unit of all living things.

All living things are made up of a cell or cells.

Cell contains information for its structural and functional development in its nucleic acids. This information is passed down from parent to offspring cells.

74E.

1. Robert Hooke
2. Mathias Schleiden
3. Theodor Schwann
4. Dujardin
5. Rudolf Virchow

74F.

A. Cell membrane

It regulates the movement of substances in and out of the cell.

It protects the cytoplasm.

It delimits the content of the cytoplasm.

C. Centriole

They help in the formation of cilia and flagella.

They provide spindle fibers to which chromosomes are attached during cell division.

D. Nucleus

The nucleus controls directly or indirectly most of the activities of a living cell.

The nucleus carries chromosomes in which hereditary materials (genes) are coded.

The DNA in chromosomes gives information for the manufacture of the proteins in the cell.

The nucleolus in the nucleus produces several kinds of RNA which are passed out of the nucleus to the cytoplasm to manufacture proteins.

E. Ribosomes

They make proteins by joining amino acids together.

F. Lysosome

Enzymes released by lysosomes destroy bacteria and cells.

They destroy worn-out parts of cells by discharging enzymes into them and thereby clearing the area for a new healthy cell to grow.

J. Golgi apparatus

May help in the manufacture of lysosomes.

May help to distribute proteins made by the cell.

May help in the formation of membranes of endoplasmic reticulum and production of cellulose of cells of plants.

K. Chloroplast

It is the seat of photosynthesis where organic foods/sugars are synthesized.

M. (Endoplasmic reticulum)

Provide surface for the location of ribosomes.

They interconnect the organelles of the cell.

They assist in the formation of nuclear membrane during nuclear division.

Transport metabolic products within cytoplasm or between the cytoplasm and nucleus.

Help in the formation of enzymes and protein.

N. Vacuole

Stores nutrients and waste products.

The cell sap is osmotic in function.

O. Mitochondrion

Centre of cellular respiration in which food substances are oxidized to release energy for the activities of the cells.

Contain enzymes and DNA (deoxyribonucleic acid).

The enzymes carry out oxidative phosphorylation of adenosine diphosphate (ADP) to adenosine triphosphate (ATP).

The DNA helps to code the synthesis of protein in mitochondrion membranes.

75A.

i. Red blood cells/red blood corpuscles/erythrocytes
White blood cells/white blood corpuscles/leucocytes.
Platelets/thrombocytes.

ii. Plasma

iii. Universal donor is blood group O universal recipient is blood group AB.

iv. Reasons: blood group O has no antigen but has antibodies. (A and B)

Universal recipient blood group AB has antigens (A and B) and has no antibodies.

75B. Transport oxygen

Temperature regulation/distributes heat to all parts of the body.

Carries digested/dissolved food/amino acids/glucose/lipids/ fatty acids and glycerol.

Carries waste products/urea/carbon dioxide.

Carries and distributes hormones.

Defends body against bacteria/engulfs bacteria/germs/ pathogens.

Produces antibodies.

Aids blood clotting.

Transports water.

75C.

Recipient blood group	Indicate the antibodies present in the spaces	Donor's blood group			
		A	B	AB	O
A	b	✓	X	X	✓
B	a	X	✓	X	✓
AB	None	✓	✓		✓
O	ab	X	X	X	✓

76A. The structure of villus

76B. Small intestine

76C. Enzymes substrates act upon

Enzymes	Substrates act upon
1. Amylase	Starch
2. Maltase	Maltose
3. Sucrase	Sucrose
4. Lactase	Lactose
5. Peptidases	Polypeptides
6. Erepsin	Peptones
7. Lipase	Fats and Oils

76E. Enzymes speed up chemical reactions/act as catalysts;

are specific in action;
are not used up or changed during reaction;
actions of enzymes are reversible;
are affected by temperature;
are affected by pH of their surrounding;
some enzymes act best in the presence of coenzymes;
enzymes can function outside organisms producing them;
and are protein in nature.
Enzymes have active sites.
Small amount of enzymes catalyze large amounts of substrate.

76F.

A. Lacteal duct
B. Epithelium
C. Vein
D. Lymphatic vessel
    (leading to vein)

E. Artery
F. Blood capillary

76G. It is long and coiled around the abdominal cavity, thus giving it a large surface area for absorption of digested food.

The walls of the small intestine are thrown into finger-like folds called villi. These villi are numerous and afford a large surface area for absorption of food.

The villi sway constantly over the intestinal content. This action makes them to be in touch with digested food.

The epithelial cells of the villi have brushlike structures called microvilli for absorption of digested food.

76H.

A. Fatty acids and glycerol
B. Amino acids and glucose
C. Amino acids and glucose

77A.

Cell		Shape	Location	Functions
I.	Sperm cell of man/ spermatozoa	Needlelike or whiplike	Testes	For reproduction
II.	Egg or ovum of woman	Round	Ovary	For reproduction
III.	Muscle cell	Spindle-shaped	Muscle	Movement/locomotion/ give shape to bodies
IV.	Nerve Cell/ neuron	Indefinite-shape	Brain and spi-nal cord	Conduction of impulses/transmit impulses/messages
V.	Red Blood cell/corpuscle/ erythrocyte	Disc-shaped	Blood plasma	Carrier of hemoglobin for oxygen transportation. Perform circulatory/excretory/ respiratory functions

VI. Bone cell/ osteoblast/ osteocyte	Star-shaped (stellate)	Bone	Gives support/aids movement/gives shape to bodies
VII. White blood cell/leucocyte	Polymorphic	Blood plasma	Defense of the body against germs

77B.

A. Acrosome
B. Nucleus
C. Mitochondria
D. Tail
E. Nuclear membrane

F. Nucleus
G. Yolky cytoplasm
H. Jelly layer
I. Plasma membrane
J. Vitelline membrane

77C.

A. Acrosome: contains enzymes that help to dissolve the cell membrane of the female ovum during fertilization.
C. Mitochondria: contains respiratory enzymes that provide energy for locomotion/swimming.
D. Tail: for locomotion.

77D.

	I (Sperm cell)	II (Egg or ovum)
1.	No yolk cytoplasm.	Yolky cytoplasm is present.
2.	Very small cytoplasm is present.	Large cytoplasm is present.
3.	Millions of sperms are released during each ejaculation.	Only one egg, or rarely two eggs, is/are released monthly.
4.	It is capable of swimming from vagina to the fallopian tube.	Not capable of swimming. It is moved by the beating of fallopian tube cilia and muscular contraction of the fallopian tube.

5.	Divided into head, neck, middle piece, and tail.	Round and not divided into head, middle piece, and tail.
6.	Very small, with a diameter of 2.5 microns.	Far larger than sperm with a diameter of 120 microns.
7.	Presence of head.	Head is not present.
8.	Tail or flagellum is present.	Tail or flagellum is absent.
9.	Absence of vitelline membrane.	Presence of vitelline membrane.

78A.

I. Receptacle
II. Sepal
III. Anther
IV. Stigma
V. Style
VI. Petal
VII. Ovule
VIII. Flower stalk/pedicel

78B.

I. IV, V, and VII
II. VI

78C. Calyx/sepals

Corolla/petals
Androecium/stamen/filament, anther, pollen grains
Gynoecium/pistil/carpel (stigma, style, and ovary)

78D.

Wind-pollinated flowers	Insect-pollinated flowers
Flowers small/inconspicuous	Flower are large/conspicuous
No scent	Scented/possesses scent
No nectar	Nectar is present
Have no particular shape	Flowers are specially shaped to facilitate pollination by particular types of insects
Flowers not brightly colored/dull	Flowers are usually brightly colored
Produce large quantity of pollen grains	Produced relatively small quantity of pollen grains
Smooth pollen grains	Rough/spiky pollen grains
Pollens grains not sticky	Pollen grains are sticky
Light pollen grains	Pollen grains are not light/are heavy
Stigmas are large/feathery	Stigma are flat/have small surface area

78F. Two examples of wind-pollinated flowers: maize, millet, rice, oat, grass.

Two examples of insect-pollinated flowers: *Delonix regia*, sunflower, pride of Barbados, hibiscus, mango, balsam.

78G. Flower: a flower is that part of the shoot modified for sexual production.

78H.

I. Receptable: carries and holds together the other parts of the flower.
II. Sepal: encloses and protects the other floral parts when the flower is in the bud stage. If brightly colored, they

also attract insects; if green, they make plant food (photosynthesis).

III. Anther: contains the pollen grains.

IV. Stigma: receives pollen grains at pollination.

V. Style: connects the stigma to the ovary, and it is the passage for the pollen tube to reach the ovules.

VI. Petal: attracts insects which pollinate flower.

79A. Transverse section of the spinal cord.

79B.

Reflex action	Voluntary action
1. Action is started by muscle receptor cells.	Action is started in the brain.
2. It is an automatic action/not under the control of will.	It is not an automatic action/under the control of will.
3. It is inborn/instinctive/innate.	It is learned.
4. It is very fast.	It is relatively slow.
5. It happens without thinking about it.	It happens after one has thought about it.
6. It involves few neurons.	It involves numerous neurons.
7. Action involves spinal cord.	Action involves the brain.

79D.

Somatic nervous system	Autonomic nervous system
1. Impulses speed along motor fibers that extend from CNS to effectors without synapses (no ganglia).	Impulses speed along motor fibers that extend from CNS to ganglia (where they synapse) and from ganglia to effectors.

2. It affects skeletal muscles.	It affects glands, cardiac muscles, and smooth muscles.
3. It always stimulates effectors.	It may stimulate or inhibit effectors.
4. Body activities are mainly voluntary.	Activities are mainly involuntary.

79F. Functions of sympathetic nervous system:

1. Accelerates heartbeat.
2. Constricts arteries.
3. Dilates the bronchioles.
4. Dilates the iris.
5. Causes relaxation of bladder muscle.
6. It slows gut movement.
7. It raises the blood pressure.
8. It inhibits the secretion of salivary glands.

Functions of parasympathetic nervous system:

1. Slows down heartbeat.
2. Dilates arteries.
3. Constricts the bronchioles.
4. Constricts the iris.
5. Causes the contraction of bladder muscle.
6. Speeds up gut movement.
7. Lowers the blood pressure.
8. It stimulates the secretion of salivary glands.

79G. Function of the spinal cord

1. It is the seat of reflex (involuntary) actions.
2. It sends impulses (message) to the brain and carries responses from the brain to the muscles

79H. Examples of reflex (involuntary actions):

Beating of the heart.	Walking
Blinking of the eyes.	Eating
Jerking of the knee.	Reading
Peristalsis.	Driving
Withdrawal of hand	Running
from a hot object.	Typing
Salivation.	Writing
Coughing.	Singing
Secretion from glands.	Dancing
Examples of voluntary	
actions:	

79I.

  A. Dorsal side
  B. Gray matter
  C. White matter
  D. Covering membrane
  E. Ventral side
  F. Central canal

80A.

  A. Incisor
  B. Canine
  C. Premolar
  D. Molars

80B. Incisor shape: rectangular, flat, chisel-like.

Function: for cutting.
Canine shape: sharp, pointed.
Function: cutting, biting, and tearing.
Premolar shape: broad, cusped (two cusps).

Function: grinding and chewing.

Molar shape: broad, cusped (four or five cusps).

Function: grinding and chewing.

80C. Dentition: refers to the number, arrangement, and conformation of teeth in an organism.

80D. Homodont: e.g., fishes; amphibians (frog, toads); reptiles (lizard).

Heterodont: e.g., rabbits, dog, man, cattle.

80E. Incisor: small, pointed, and chisel-shaped/sharp.

Used for cutting/tearing flesh from bone, or holding pieces of food, or used for offence.

Canines: long, strong, curved, sharp, and pointed.

Used for killing prey.

Tearing flesh from bone.

For holding prey from escaping.

Premolars: have flat surfaces.

Pointed with sharp cusps on the edges.

For cutting large pieces of meat.

Last pair of upper premolar and one pair of lower molar form carnassial teeth.

Used for tearing flesh from bones and for breaking/grinding bones.

Molars: are cusped.

Used for crushing/breaking bones.

Grinding bone and flesh/meat.

80F.

i)   Man: $\dfrac{I\,2,}{2}$ $\dfrac{c1,}{1}$ $\dfrac{p2,}{2}$ $\dfrac{m3}{3}$

ii)   Dog: $\dfrac{i3,}{3}$ $\dfrac{c1,}{1}$ $\dfrac{p4,}{4}$ $\dfrac{m2}{3}$

iii)   Rabbit: $\dfrac{i2,}{1}$ $\dfrac{c0,}{0}$ $\dfrac{p3,}{2}$ $\dfrac{m3}{3}$

iv) Horse: $\dfrac{i3,}{3}$ $\dfrac{c1,}{1}$ $\dfrac{p4,}{4}$ $\dfrac{m3}{3}$

81A. Vertical section through the human ear

81B.

    i.   Hearing
    ii. Balancing

81C.

    i.   The outer ear/pinna
    ii.   Middle ear
    iii. Inner ear (labyrinth)

81D.

    A. Pinna
    B. Auditory canal (meatus)/ear tube
    C. Ear drum/tympanic membrane

D. Eustachian tube
E. Semicircular canals
F. Auditory nerve
G. Cochlea

81E.

A. Pinna: pinna collects soundwaves and directs them into the auditory meatus.
Protects the internal structures.
It detects the directions of soundwaves.
B. Auditory canal or meatus
It prevents the entry of tiny insects, germs, and dust.
Receives soundwaves from the pinna/outer ear.
C. Eardrum or tympanic membrane
Helps to transmit soundwaves from the outer ear to the middle ear.
D. Eustachian tube
The tube allows air from the surroundings to enter or leave the middle ear so that the air pressure on both sides of the eardrum is equal.
G. Cochlea
Contains nerve cells sensitive to sound vibrations and, therefore, concerned with hearing.

81F.

I. Mechanism of hearing:
The pinna collects soundwaves in the air, concentrates them, and passes them on through the external auditory canal meatus; the waves cause the tympanic membrane/ eardrum to vibrate.

The vibrations are passed onto the ear ossicles which amplify them.

The round oval window/fenestra ovalis/rotunda also vibrates, passing the waves into the cavity of the inner ear/cochleas where the perilymph vibrates/transmits the waves into the cochlea, causing the endolymph of the cochlea to vibrate.

The vibrations are transmitted across the organ of sensory nerve cell/basila membrane.

Impulses are setup which stimulates the auditory nerve cells, which then transmit the impulses to the brain for interpretation.

II. Mechanism of balancing:
Head movement in any direction affects the fluids/endolymph in the corresponding semicircular canal which are at right angles to each other, forcing the sensory cells/gelatinous cupula in the ampulla to setup impulses through the auditory nerve to set the brain and, in turn, for interpretation.

The brain relays impulses to the body muscles.

81C.

i.   Endolymph
ii. Perilymph

82A.

I.   Longitudinal/vertical section of coconut/drupe
II.   Longitudinal/vertical section of tomato/berry
III. Fruit of desmodium
IV.   Pod/legume of *Caesalpinia pulcherima* (pride of Barbados)
V.   Capsule of Dutchman's pipe

82B.

    I.   Drupe
    II. Berry
    III. Schizocarp
    IV. Legume/pod
    V. Capsule

82C.

    I.   Coconut/drupe
        Has a thin epicarp.
        Has a fibrous mesocarp.
        Has hard/stony endocarp.
        Has only one seed.
    II. Tomato/berry
        Thin/membranous epicarp.
        Fleshy/succulent mesocarp/endocarp.
        Mesocarp and endocarp are fused.
        Has many seeds.
    III. Desmodium/schizocarp
        Has one carpel.
        Syncarpous ovary with many seeds.
        Fruits breaks into units, each enclosing one seed.
        Has hairs with hooks.
    IV. Legume/pod/pride of Barbados
        Formed from one carpel.
        Seeds arranged along a margin.
        Split along both sides of the pod.
        Long and flattened.
    V. Capsule/Dutchman's pipe
        Several seeds.
        Splits along several longitudinal slits.
        Many carpels/polycarpellary.
        Carpels fused/syncarpous.

82D. Agent of dispersal of specimen C/desmodium

    i.   Animal/man
    ii. Has sticky/tiny hairs/hooks which stick to the hair/clothing of animal/man.

82E. To avoid overcrowding.

    To reduce competition (for food, water, and light).
    To spread plants to different/new areas.
    To increase survival of species.

82F.

A.	Epicarp	G.	Endocarp
B.	Mesocarp	H.	Seed
C.	Endocarp	I.	Epicarp
D.	Seed	J.	Mesocarp
E.	Funicle	K.	Persistent calyx
F.	Fruit stalk		

83A. Manometer/manometer with cut root/instrument/apparatus for measuring root pressure

83B. Measurement of pressure/root pressure

83C. Mercury

83D. From I

84A.

    I.   Femur of rabbit/rat/small mammal
    II. Humerus of rabbit/rat/small mammal

84B.

A. Head of humerus
B. Shaft of humerus
C. Olecranon notch
D. Condyle
E. Trochlea
F. Supratrochlear notch
G. Deltoid ridge

H. Groove for patella
I. Shaft
J. Lesser trochanter
K. Neck
L. Head
M. Condyle
N. Greater trochanter

84C.

Diagram I (Femur)	Diagram II (Humerus)
Longer	Relatively shorter
Head is prominent	Head is not prominent
Projections are prominent	Projections are not prominent
No hole above trochlea	Hole is above trochlea
Distal end knobbed/ not grooved	Surface of distal end grooved
Distal end curved toward posterior	Not curved at distal end
Head is with short neck	Head is without neck

84D. Function of I (femur)

Forms attachment surface for the thigh muscles.

Forms a ball and socket joint with the ilium.

Allowing the free movement of the hind limb.

The distal end forms a hinge joint with the patella, tibia, and fibula.

Functions of II (humerus)

Allows free movement of the arm.

Forms the hinge joint at the elbow.

Forms attachment surface for biceps and triceps muscles.

Forms the ball-and-socket joint at the shoulder.

The bicipital groove anchors the biceps muscles.

84E.

84F. Bone that forms joint with II/humerus at proximal end is scapula.

84G. At distal end are ulna and radius.

84H.

    i.   Type of joint at proximal end is ball and socket.
    ii. Joint at the distal end is hinge joint.

85A.

    I.   Butterfly
    II. Grasshopper or cockroach

85B.

A. Labial palp	F. Antenna
B. Mandible	G. Labrum (upper jaw)
C. Maxillary palp	H. Labium (lower jaw)/
D. Proboscis	second maxillae
E. Compound eye	I. Labial palp

85C.

    I.   Sucking mechanism

    II. Chewing (biting) mechanism

85D.

Mosquito: sucking mechanism
Housefly: sucking mechanism
Tapeworm: absorbing mechanism

85E.

    G. Labrum (upper lip): prevents the food from falling off the mouth.

    H. Labium (lower lip): prevents wastage of food from the mouth.

85F. Mandible

86A.

    i.   Glass tube trachea/windpipe

    ii. Bell jar: rib cage/thoracic cavity/chest/thorax

    iii. Branches of glass tube: bronchi

    iv. Balloons: lungs

    v. Polythene or rubber membrane: diaphragm

86B. Behavior of the balloons

    i.   When the rubber membrane is relaxed:
        The pressure in the bell jar increases;
        volume within bell jar decreases; and
        air is expelled from the balloons through the glass tube.

    ii. When the rubber membrane is pulled down:
        Pressure in the bell jar decreases;

volume within bell jar increases; and
air enters in to the balloons through the glass tube.

iii. Inspiration in mammals:
Contraction of the intercoastal muscles/diaphragm, lowers the diaphragm, and raises the rib cage.

These bring about an increase in volume of the thoracic cavity and lowers the pressure.

Air therefore rushes into the lungs, represented by the balloons, through the nostrils and into the trachea (represented by the glass tube).

87A.

i.
- A. Epigeal
- B. Hypogeal

ii.
- A. Cowpea, groundnut, melon, mango, beans, etc.
- B. Maize, guinea corn, millet, wheat, etc.

iii.
- A. Foliage leaf
- B. Epicotyl
- C. Hypocotyl
- D. Tap root
- E. Plumule
- F. Cotyledon
- G. Radicle
- H. Coleoptile
- I. Fibrous root
- J. Radicle

iv. Water or moisture
Air or oxygen
Warmth or suitable temperature
Seed must be viable

Enzymes

v.

    A.   Tap root

    B.   Fibrous root

vi. A is a dicotyledon

    B is a monocotyledon

vii.

Dicotyledon	Monocotyledon
Seed has two cotyledons	Seed has one cotyledon
Taproot system	Fibrous root system
Secondary growth is usual	Rarely secondary growth
Net-veined leaves	Parallel-veined leaves, but the yam is an exception
Vascular bundles arranged in rings in the stem	Vascular bundles arranged irregular in the stem
Flowers in fours, fives, or multiple of these	Flowers in threes and multiples of threes

88A.

    A.   Scolex/head of tapeworm (*Taenia solium*)

    B.   Mouthpart of housefly

88B.

    A.   Rostellum

    B.   Hooks

    C.   Sucker

    D.   Neck

    E.   Young proglottids

    F.   Compound eye

    G.   Antenna

    H.   Palp

    I.   Proboscis

88C.

    A. Absorbing mechanism
    B. Sucking mechanism

88D. Sponging

88E. A, B, C, (rostellum, hooks, and suckers) are used by the organism to fasten itself to the lining of the host's intestine, or for attachment to the host.

89A.

    i.   Feathers
    ii.
        I.   Quill feather
        II.   Covert feather/contour
        III. Down feather
        IV. Filoplume feather
    iii.
        A. Vane
        B. Shaft/rachis
        C. Aftershaft
        D. Superior umbilicus
        E. Quill
        F. Inferior umbilicus
    iv. Birds
        Class: Aves
        Phylum: Chordata
    v.   Functions of the feathers:
        Insulation/regulation of body temperature: incubation
        Courtship display: flight
        Protection: sex identification
        To give shape

90A.

   i.

      I.   Bird or any named bird/dove, chicken, pigeon, etc.

      II.   Egg (bird's egg/transverse section of bird's egg)

   ii. Bird lays (II) egg.

   iii.

      A.   Nostril

      B.   Claw

      C.   Eye

      D.   Beak

      E.   Wing

      F.   Flight/quill feather

      G.   Tail feather

      H.   Shell

      I.   Air space

      J.   Albumen

      K.   Germinal disc (embryo)

      L.   Yolk

      M.   Chalaza

   iv. Functions

      D.   Beak: for picking food/cracking seeds/for feeding.

      H.   Shell: protect the egg/it is porous, hence aids respiration for the embryo.

      I.   Air space: for respiration of the embryo.

      J.   Albumen: rich in protein for nourishment of the embryo.

      L.   Yolk: supplies nutrients to the embryo.

      M.   Chalaza: holds the yolk and embryo in position within the albumen.

   v.

   A.

      1.   Quill feather

      2.   Contour or covert feather

      3.   Down feather

      4.   Filoplumes

      5.   Bristle feather

B.
1. Quill: found on the wings and tail.
2. Contour or covert: cover the body.
3. Down feather: found between the contour feathers.
4. Filoplumes: found scattered all over the body.
5. Bristle feather: found around the eyes, nostrils, and on the head.

C. Quill: for flying and steering.
Contour or covert feather: helps to keep the body warm.

Down feather: helps to keep the bird warm.

Filoplumes: found scattered all over the body; they have unknown functions (may insulate the body).

Bristle feather: helps to prevent foreign bodies from the eyes and nostrils.

vi. Class: Aves

Characteristics of the Class (Aves)

1. They are homoiothermic or warm-blooded animals/have a constant body temperature.
2. Bodies are covered with feathers except the hindlegs, which are covered with scales.
3. They have two pairs of limbs.
4. They have wings which are used for flight.
5. They have beak which is used for feeding.
6. They have rigid and hollow bones with air sacs which make them light during flight.
7. They have a four-chambered heart.
8. Their reproduction is sexual, and fertilization is internal.
9. They exhibit oviparous mode of reproduction.
10. They have lungs which are used for gaseous exchange.
11. They show parental care for the young ones.

91A. A profile of a (tropical) rainforest:

91B. Upperlayer/emergents

Middle layer
Lower layer (canopy)
Shrub layer
Ground layer/flora (forest floor)

91C. Vegetation consists mainly of woody plants

Rainfall is abundant
Moderate temperature
Very rich in both plant and animal species
Trees are stratified/trees are in layer/canopies
Climbers/epiphytes are present
Some of the trees have well-developed buttress roots
Tall trees
Trees with broad leaves/presence of canopies
High relative humidity

91D. Rainfall: relative humidity

Temperature: sunlight
Wind

91E. Trees

African walnut
Mahogany
Opepe
Obeche
Iroko
Animals
Pangolin
Bush cow

Elephant
Snail
Monkeys
Snakes
Chameleon
Birds

91F. Tropical rainforest

Guinea savanna
Sudan savanna
Sahel/sahelian savanna
Montane forest
Mangrove/brackish water/estuarine swamp forest

92A.

B. Shortsightedness/myopia
C. Longsightedness/hypermetropia

92B.

i. Shortsightedness/myopia can be corrected by using concave lens eyeglasses.
Longsightedness/hypermetropia can be corrected by using convex lens eyeglasses.

ii. Effects of the correction
In B/Shortsightedness: the concave lens will diverge the light and cause image to form on the retina.
In C/Longsightedness/hypermetropia: the convex lens will converge the light rays and cause images to form on the retina.

92C. Accommodation: the ability to focus images on the retina is called accommodation/ability of the eye to focus (both near and distant).

Objects on the retina:

92D. Vertical section (longitudinal) of the human eye

92E.

1. Aqueous humor
2. Conjunctiva
3. Cornea
4. Pupil
5. Iris
6. Lens
7. Suspensory ligament
8. Ciliary muscle

9. Vitreous humor
10. Sclerotic coat/sclera
11. Choroid coat
12. Retina
13. Yellow spot
14. Blind spot
15. Optic nerve

92F. Aqueous humor/I

It refracts light rays into the retina.

It helps to maintain the spherical shape of the eye iris/5.

Controls the amount of light passing through the eye.

Sclerotic coat/10

Gives shape and firmness to the eye.

Protects and supports the inner parts of the eye.

Retina/12

Images are formed on the retina.

Helps to detect colors of object.

Light rays come to a focus on retina.

Yellow spot/13

It is the point where image is focused.

It is the most sensitive part of the retina.

Fullest visual information is sent to the brain through the yellow spot.

Optic nerve/15

Transmit sensory impulses to the brain.

Transmit sensory impulses from the brain.

92G. Similarities:

Lens is converging in nature.

Both retina and film are sensitive to light.

Both iris and diaphragm are used to control the amount of light.

The image formed is real, inverted, and diminished.

Differences

Human Eye	Camera
1. Its focal length varies	Focal length is fixed

2. It records the image into a pattern of electrical impulses which are sent to the brain through the optic nerve	It records the image in light-sensitive silver salt crystals on the film
3. Human eye adjust focal length by contraction or relaxation of the ciliary muscles	Objects are focused by mere shifting of the lens forward or backward

93A. Skull of herbivore/herbivorous animal/rabbit/guinea pig/rat

93B. Kingdom: Animalia

Phylum: Chordata/Vertebrata
Class: Mammalia

93C.

I.   Cranium/brainbox
II. Lower jaw of
      a herbivore
III. Molars/molar teeth
IV. Diastema

V.   Incisor
VI. Upper jaw of
      herbivore
VII. Orbit/orbital cavity/eye socket

93D. Digestive/alimentary canal

93E. Herbivorous mode of feeding/herbivore

93F. Rabbit, squirrel, guinea pig, grass cutter, rat, mouse

93G.

III. Lower molars: flat surface used for grinding.

IV. Diastema: a gap that allows the tongue to move food around the mouth for effective grinding.
V. Incisor: have chisel edges for cutting.

93H. Differences

Structure illustrated	Skull of a carnivore
Incisors well-developed	Incisors are less developed
Carnassial teeth are absent	Carnassial teeth are present
Canines are absent	Canines are present
Diastemas are present	and prominent
Crowns of molar flattened	Diastemas are absent
	Sharp cusps in molars

Similarities:
Incisors are present.
Premolars are present.
Molars are present.
Presence of fused/fixed sutures.
Presence of upper jaw.
Presence of lower jaw.

94A.

I. A potometer
II. To measure transpiration rates of a plant shoot under different environmental conditions
III.
    A. Leafy shoot
    B. Air bubble
    C. Screw clip
    D. Rubber tubing
    E. Capillary tube
    F. Scale
    G. Support

IV. Precautions

1. The shoot to be used must be cut underwater to prevent the entry of air which will likely block up the xylem vessels in the stem and disturb the movement of water through them.

2. The whole apparatus is carefully filled with water. There should be no airspace in it.

3. When the shoot is quickly introduced through the rubber cork, it is immediately greased with Vaseline to keep the apparatus airtight.

GRAPH

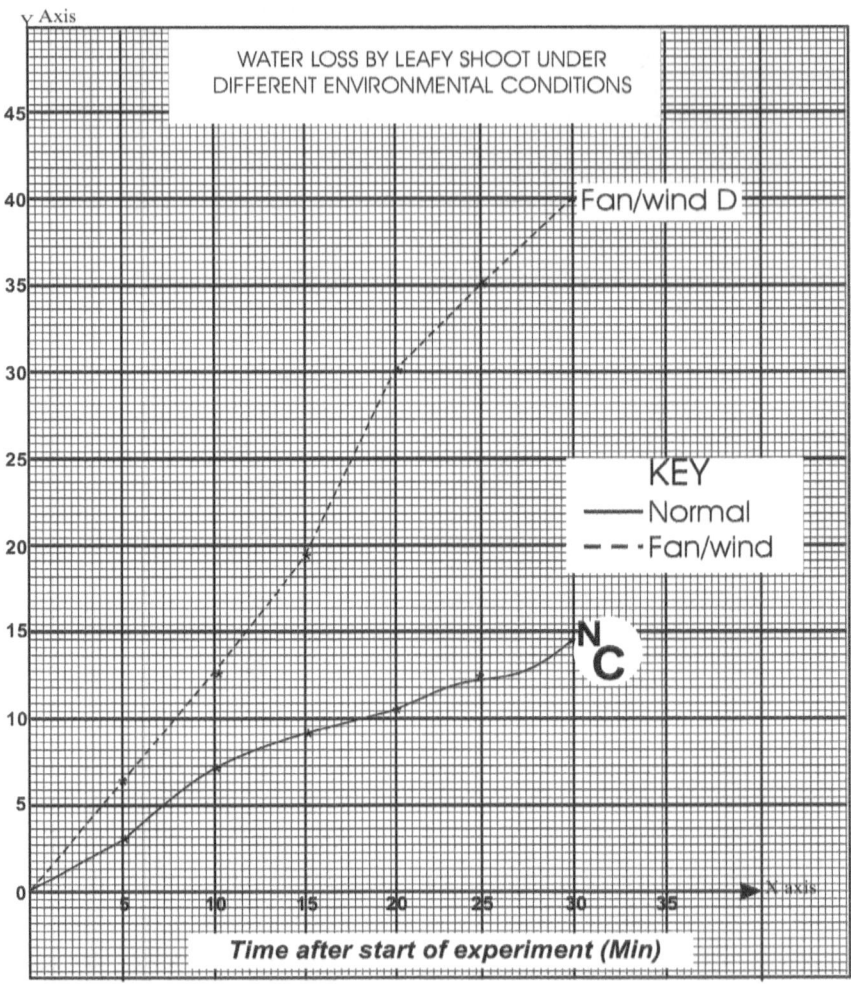

94B.

II. Time interval of five minutes.
III. Under the fan/wind/windy/breeze/breezy condition/condition D.
IV. Effect of fan:
Wind/breeze removed water vapor above the stomata, creating a greater moisture gradient/difference between the levels of internal and external environments/atmospheres.

Making the water vapor move faster out of the leaf/increased transpiration/evaporation.

Therefore, water is absorbed faster by the plant shoot/root of shoot to replace the lost water.

V. To study the effect of environmental conditions/wind/breeze/moving air; on the rate of water loss/transpiration/uptake of water by shoots.
VI. Transpiration
VII. Stomata (sunken): thick cuticles
Guard cells: thin cell wall
Chloroplast/chlorophyll
Compact mesophyll/palisade layer

95A.

1. Boiling the leaf in already boiling water for five minutes.
2. Washing the leaf in hot water.
3. Boiling the leaf in 70% alcohol (ethanol).
4. Adding few drops of iodine solution.

95B.

1. Boiling the leaf in already boiling water for five minutes.
3. Boiling the leaf in 70% alcohol (ethanol).

2. Washing the leaf in hot water.
4. Adding few drops of iodine solution.

95C.

1. Boiling the leaf in water for five minutes.
   Reason: boiling kills the protoplasm and makes the cell membrane permeable.
   Stops all enzyme activities in the cells.
   Makes iodine more easily permeable to starch granules at the time of test.
2. Washing the leaf in hot water.
   -Reason: to soften the leaf.
3. Boiling the leaf in 70% alcohol.
   -Reason: to extract chlorophyll from the leaf/decolorize the leaf.
4. Adding few drops of iodine solution.
   Reason: to test for the presence of starch.

95D.

A. Boiling tube
B. Water bath/beaker
C. Hot water
D. Boiling alcohol
E. Blue-black color

96A.

V. General structure of a bacterium/bacterial cell
VI.
   A. Slime capsule
   B. Cell wall
   C. Glycogen granules
   D. Cytoplasm
   E. Nuclear material

F.   Flagella
VII.
    A.   Flagellated bacilli
    B.   Streptococci
    C.   Staphylococci
    D.   Vibrios
    E.   Spirochaetes
VIII. Diseases
    B. Flagellated bacilli: typhoid
    C. Streptococci: sore throat
    D. Staphylococci: boil
    E. Vibrios: cholera
    F. Spirochaetes: syphilis
IX.   Diseases of plants caused by diagram A.
    Soft rot of tomatoes, soft rot of carrot, ring rot of pota-
        toes, sliminess of vegetables, potato leaf blight, web
        blight of cowpea, crown gall of apple, cassava blight,
        coconut blight, cotton blight.
    Bacterial wilt of plants, fire blight of apple.
    Diseases of animals caused by diagram A.
    Gonorrhea, tuberculosis, tetanus lockjaw, whooping cough,
        leprosy, pneumonia, anthrax, dysentery, mastitis.
X.   Importance
    Manufacture of cheese/yogurt/butter
    Fermentation of alcohol and beverages
    Nitrification by legumes and symbiotic bacteria
    Manufacture of vaccines
    Aid in digestion in herbivores and termites
    Causes food spoilage/decay of organic matter
XI.
    1.   Diagram A (bacterial cell/bacterium) has no nuclear
          membrane; the nuclear material diffuses through the
          cytoplasm.
    2.   The cytoplasm of diagram A is very dense, con-
          tains ribosomes, but has no mitochondria nor Golgi
          bodies.

3. Diagram A has a cell wall, but this is not made of cellulose.

XII. In air, soil, dust, freshwater, sea, dead plants, animals, and inorganic materials.

97A.

Diagram	Type of bird	Type of feet	Special uses
1.	Owl	Long, sharp claws	For catching and holding prey
2.	Domestic fowl	Strong feet but blunt nails	For scratching in the earth
3.	Duck	Webbed between toes	For swimming quickly

97B.

Type of bird	Type of beak	Special uses
Eagle	Hooked	For killing prey and tearing off strips of flesh
Parrot	Hooked	For cracking seeds but is also used in climbing
Swallow	Short with wide gap	For catching insects in flight
Ibis	Long, narrow, and curved	For probing earth and mud

98A.

i. Paramecium

ii.

    I. Cilium

    II. Contractile vacuole (anterior)

    III. Pellicle

IV. Food vacuole
V. Ectoplasm
VI. Endoplasm
VII. Gullet/oral groove/mouth
VIII. Contractile vacuole (posterior)

98B.

I. Cilium for locomotion/movement/swimming
II. Contractile vacuole for osmoregulation/removal of excess water/excretion
III. Pellicle to maintain body shape/allows water entry/protects the animal
IV. Food vacuole for digestion of food materials

98C. There will be entry of water into the cell/animal:

leading to rapid formation of contractile vacuoles; and
to remove excess water entering the organism

98D.

I. Cell/cellular level
II. The single cell performs all functions of a living organism

98E. By simple diffusion, waste products (i.e., water/ammonia/carbon dioxide).

Enter into the canals of the star-shaped contractile vacuoles at both ends of the organism when the canals are full.

Contraction of the canals force the contents of the canals into the center of the contractile vacuole, from where the wastes are squashed out of the cell into the surrounding water alternately.

99A.

S_1 = first sample collected = 250
S_2 = second sample collected = 250
S_3 = marked individuals in the = 50

Second sample
Total population = $\dfrac{S1 \times S2}{S_3}$ = $\dfrac{250 \times 250}{50}$ = 1,250 individuals

99B. Precautions:

Area under investigation should be so controlled to prevent migration/emigration and immigration of animals.

Marking should be "unwashable"/indelible.

Markings should not attract predators.

Markings must not impede the movement of the marked animals.

Handle captured organisms with care.

99C. Uses of line transect:

A chain/string is marked/knotted at regular intervals and laid across the area under investigation.

Intervals should be 50 cm to 100 cm.

Members of a particular plant species touching the chain at marked/knotted points are counted (X).

All plants which touch the chain at the marked/knotted points are counted (Y).

Layout several transects and repeat the counting of the particular species as well as the total number of plants as done above.

The percentage of that particular plant species is then calculated as $\frac{x}{y} \times \frac{100}{1}$

99D. Determination of density of plant species in an abandoned farmland:

Select a suitable plant in an abandoned farmland.

Determine the size of the quadrat to be used.

Measure the area under study.

Randomly toss the quadrat backward over the shoulder to avoid bias.

Several tosses/at least ten tosses should be made.

Each time the number of species within the quadrat is counted and recorded, the average number of the species of the total tosses is determined.

Divide the average number of times the species occurs within the quadrat by the area of the habitat.

This gives the density of the species.

100A.

    I.
        i.   Lizard/*Agama*
        ii. Toad/*Bufo regularis*
    II.
        A. Reptilia chordata

B. Amphibia chordata
III. Scales: nostrils
Limbs
IV. Significance
Scales: prevent desiccation for survival on land.
Limbs: limbs were necessary for movement on land.
Nostrils: necessary for breathing on land.
V. Adaptive features of limbs
Digits/end in claws for digging/gripping/climbing
Limbs are long and bent outward at an angle for raising
the body and for movement.
Hind limbs longer than forelimbs.
Forelimbs shorter for jumping and landing.
VI.
A. Scale
B. Eye
C. Nostril
D. Mouth
E. Eardrum
F. Gular fold
G. Neck
H. Forelimb
I. Digit with claw
J. Trunk
K. Hind limb
L. Tail
VII. Male features:
Orange/red head
Tail with blue/gray color
Prominent nuchal crest
Cloacal opening bordered by scales/pre-anal pads
Conspicuous gular fold
Presence of hemipenis
Female features:
Greenish/brownish/grayish/black head
Color of body: brown/grayish black with yellow spots

Reduced/less prominent/small nuchal crest

Less prominent gular fold

Cloacal opening not bordered by scales/pre-anal pads

Absence of hemipenis

VIII. Territorial behavior in Agama lizard:

In a mature Agama lizard's territory, there is usually one adult male, several mature females, and juvenile (young) males and females.

No adult male is allowed.

If an adult male intrudes into the territory, the resident male that owns the territory threatens the intruder by head nods and threatening postures.

The adult male can also expand its gular fold and scare off the other male lizard.

If these fail to produce the desired result, then the adult male engages in a fight in which the tails are used to whip each other.

100B.

I.

A. Poison gland

B. Nostril

C. Eye

D. Mouth

E. Eardrum/tympanum

F. Digit

G. Forelimb

H. (Swollen) abdomen

I. Webbed digit/Hindlimb

J. Anus

II.

Forelimb	Hind Limb
Short	Elongated/long
Less muscular	More muscular
Bears four digits	Bears five digits
Digits not webbed	Webbed digits

IV. Importance of the differences:
Short/stout forelimb: takes the shock of landing/for land-ing/for support.

Muscular hindlimb: enables takeoff when hopping.

Webbed digits of hindlimb for paddling in water/for swimming

V. Near a stream/pond/river/under (wet) log of wood/stone/ in cool/moist/damp places

VI. Frogs are more slightly built than toads (diagram B), and the body is somewhat pointed at both ends.
The skin of a frog is smooth and slimy, and they spend more time in water than diagram B (toad).

The toes of the frog are joined by a thin web of skin throughout their length which is absent in the frog.

The frog is better at hopping than the toad.

Frogs have simple teeth in the upper jaw on the roof of the mouth but absent in toads.

ABOUT THE AUTHOR

The author is Adewoye Babatunde Akinwumi, who bagged BSc Ed. Biology and MSc Zoology (Parasitology option) both from University of Ilorin, Kwara state, Nigeria, in 1989 and 1992 respectively. He has taught high school biology for over twenty years, and he formerly worked with the Federal Ministry of Education, Nigeria, where he taught biology in various federal government colleges: Federal Government Girl's College, Bwari, Abuja (Federal Capital); Federal Government College, Ogbomosho, Oyo State; and Federal Science and Technical College, Usi-Ekiti, Ekiti State. He was, at various times, head of the subject in Oyo and Ekiti States.

He was for many years a WAEC (West African Examination Council) and NECO (National Examination Council) examiner.

The author is proud to put on record that the student performance in biology used to be the best compared to other subjects in the external examinations conducted by WAEC and NECO while working in those schools. The content of *Insight Biology* being used helped to achieve this feat. The author rose to the position of an assistant director of education in the Federal Ministry of Education, Nigeria, and eventually relocated out of the country at the exact time he was given a duty post to become vice principal in one of the federal government colleges in the northern part of Nigeria.

The author currently has found himself in another profession where the subject he taught for several years is one of its backbones. The author has been a member of American Society of Parasitology and a member of Phi Theta Kappa Honor Society.